Leckie
he education publisher
for Scotland

Primary **Maths**
for **Scotland**

Early Level Maths

Problem Solving Pack

Author: Sheena Dunlop
Series Consultant: Carol Lyon
Series Editor: Craig Lowther

© 2022 Leckie

001/17082002

10 9 8 7 6 5 4 3 2

ISBN 9780008508661

Published by
Leckie
An imprint of HarperCollinsPublishers
Westerhill Road, Bishopbriggs, Glasgow, G64 2QT

T: 0844 576 8126 F: 0844 576 8131

leckiescotland@harpercollins.co.uk www.leckiescotland.co.uk

HarperCollins Publishers
Macken House, 39/40 Mayor Street Upper, Dublin 1, D01 C9W8, Ireland

Publishers: Fiona McGlade and Jennifer Hall
Project editor: Peter Dennis

Special thanks
Copy editor: Louise Robb
Illustrations: Ann Paganuzzi
Layout: Siliconchips Services Ltd

A CIP Catalogue record for this book is available from the British Library.

Acknowledgements
Images © Shutterstock.com
Whilst every effort has been made to trace the copyright holders, in cases where this has been unsuccessful, or if any have inadvertently been overlooked, the Publishers would gladly receive any information enabling them to rectify any error or omission at the first opportunity.

Printed in the United Kingdom.

This book contains FSC™ certified paper and other controlled sources to ensure responsible forest management.

For more information visit: www.harpercollins.co.uk/green

Contents

Download free resources at www.collins.co.uk/primary-maths-for-scotland-free-resources

Introduction

Primary Maths for Scotland overview

Primary Maths for Scotland comprises a whole-school mathematics programme which also includes the Scottish Curriculum for Excellence First and Second Levels. The Primary Maths for Scotland series is designed to develop deep conceptual understanding and can be easily implemented alongside existing plans. It has been written by a team of Scottish Numeracy and Mathematics specialists, especially for Scottish schools, and has been aligned to the Scottish Curriculum for Excellence Benchmarks to plan routes through the Experiences and Outcomes for Numeracy and Mathematics.

Whilst this Early Level Problem Solving book complements the Early Level Primary Maths for Scotland Digital Package, Teacher Guide, Record Book and Assessment Pack, it can also be used as a standalone resource. Research informed pedagogy was used to guide the creation of these resources. The theory of social constructivism underpins the approach and has informed the phrasing of resources and teaching ideas.

Young children are naturally curious. Skilled educators nurture this curiosity by providing engaging contexts and asking rich questions which interest and challenge them to talk about their ideas and strategies for solving mathematical problems. We want all pupils to enjoy learning mathematics. This approach supports positive learning experiences and the development of metacognition.

Problem Solving

In his seminal work, Polya (1945) articulated the process of solving problems as:

1. Understand the problem
2. Devise a plan
3. Carry out the plan
4. Examine the solution obtained (look back and reflect)

Since the publication of Polya's work, our understanding of problem solving and its importance to learning, life and work has grown considerably. Problem solving is far more than a process; it incorporates thinking skills, attitudes towards learning and strategies for solving problems. As stated in the *Mathematics Principle and Practice* paper, "Mathematics is important in our everyday life, allowing us to make sense of the world around us and to manage our lives. Using mathematics enables us to model real-life situations and make connections and informed predictions. It equips us with the skills we need to interpret and analyse information, simplify and solve problems, assess risk and make informed decisions." (Scottish Government, 2011)

The ability to solve problems is vital to every child and young person as they grow and develop. In addition to supporting pupils' understanding of the curriculum, good problem-solving behaviours allow them to make sense of their world, preparing them for their future lives and for the world of work. Indeed, in many high-performing education systems, problem solving is considered the goal of the mathematics curriculum. In these systems, Problem Solving is at the heart of the curriculum, building not only on sound mathematical processes, skills and understanding of concepts, but also on collaboration, communication, metacognition and a positive attitude to learning mathematics. This Primary Maths for Scotland problem-solving resource has been written with all of the aforementioned aspects in mind.

An important part of the problem-solving process is allowing pupils space to *puzzle*, space to try out and discuss ideas, and space to be creative in their response to the problem posed. We do not advocate leading pupils through a problem, nor introducing a specific strategy in advance of the problem-solving activity. To do so can reduce pupil agency, curtail opportunities to link previous learning to this new situation and stifle creativity.

Through the process of solving problems, pupils will develop a range of transferable strategies, methods and skills. These should be reviewed and celebrated in order to scaffold the pupils' thinking and encourage metacognition.

Linking with the Numeracy and Mathematical Skills

"To face the challenges of the 21st century, each young person needs to have confidence in using mathematical skills, and Scotland needs both specialist mathematicians and a highly numerate population." (Scottish Executive, 2006)

The Benchmarks for Numeracy and Mathematics (Education Scotland, 2017) outline eight overarching skills, embedded within the Experiences and Outcomes. By following the advice given in this resource, teachers can support the development of these skills. The following notes how each skill can be developed through engagement with the Primary Maths for Scotland problem-solving tasks.

- **interpret questions.** For every problem the pupil will need to work out what is being asked of them and decide how to proceed. Some will need to ask questions to clarify their thoughts or to simplify the problem in their own mind.

- **select and communicate processes and solutions.** The pupils will need to choose a starting point and be challenged to justify their choice. As they work on the problems, they should be encouraged to share their thinking. This could be through a discussion, by sharing a picture or by using concrete materials to model or act out the situation. All of these processes are enhanced when pupils are encouraged to share and talk about their approach.

- **justify choice of strategy used.** The Let's discuss section suggests a range of possible strategies that pupils might use. In addition to sharing strategies and ideas as they work on a problem, teachers should initiate a whole group plenary in which the relative merits of the different strategies used by the class/group are discussed. By asking the pupils to draw comparisons between the methods used, and with methods that they have encountered previously, we can provoke metacognition.

- **link mathematical concepts.** Many of the problems use mathematical concepts from across the Early Level maths curriculum. Being able to link concepts, and connect different areas of maths together, will help pupils experience success and have a positive impact on their confidence.

- **use mathematical vocabulary and notation.** The problems in this pack cover almost all of the Experiences and Outcomes at Early Level. As such, there will be many opportunities for the pupils to practise using developmentally appropriate mathematical vocabulary.

- **use mental agility.** There are opportunities for the pupils to decide how to count and calculate. Some children may draw pictures or make jottings of their thinking (see Experiences and Outcomes identified against each problem).

- **reason algebraically.** There are opportunities for the pupils to look for and continue patterns, spot relationships, generalise and represent the same problem in different ways (see Experiences and Outcomes identified against each problem).

- **determine the reasonableness of a solution.** For every problem, we suggest that pupils are challenged to justify their ideas through the use of carefully selected questions, for example, Tell me about…. How do you know? Why did you choose…?

The Problems

Each problem in the resource has the following structure to help you plan and carry out the activity with your class or group.

Experiences and Outcomes

Experiences and Outcomes

- *MNU 0-02a – I have explored numbers, understand that they represent quantities, and I can use them to count, create sequences and describe order.*
- *MNU 0-03a – I use practical materials and can 'count on and back' to help me understand addition and subtraction, recording my ideas and solutions in different ways.*

Numeracy and Mathematical Skills

- **Selects and communicate processes and solutions** – shares thinking; verbalises or demonstrates thought processes
- **Justify choice of strategy used** – shows and talks through their thinking
- **Mental agility** – knowledge of number facts

We have identified the Experiences and Outcomes that are directly linked to the problem, and the overarching skills pupils may use to solve it. Pupils may well end up making links with other areas of the mathematics curriculum not identified. This is the nature of problem solving and is to be celebrated.

Resources

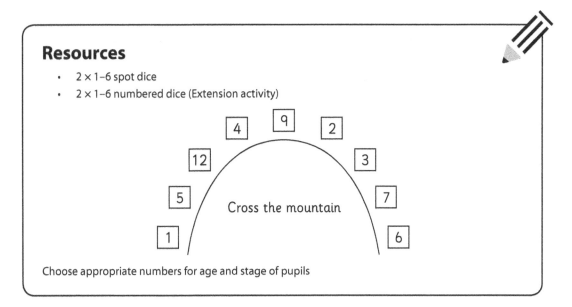

Resources

- 2 × 1–6 spot dice
- 2 × 1–6 numbered dice (Extension activity)

Cross the mountain

Choose appropriate numbers for age and stage of pupils

This section lists the physical resources, interactive whiteboard visuals and templates that should be available to support the problem-solving process. Where visuals or templates are required, we have provided these as downloadable resources (in colour).

Go to **www.collins.co.uk/primary-maths-for-scotland-free-resources** to download your free resources.

In your setting you will probably have a range of items available to the children; paper, boxes, counters, pinecones, cubes, play money, buttons, pebbles, leaves and counters to name a few. To help the children to become problem solvers we wait to see what items the children want to explore with and the approach they want to try first. Some will want to talk through the problem and ask questions. Some will act out the situation using real objects or manipulatives. Others may record their ideas by drawing a picture or mark making.

Before they start

> **Before they start...**
>
> The pupils should be able to add within 12 and subtract within 6.
>
> The pupils should be able to identify and recognise numbers up to 12

We have listed the specific skills, knowledge or processes that pupils will need to access the problem. The pupils should be familiar with these and, if possible, the context of the problem before they start. This frees up cognitive space for the pupil to focus on the problem in hand. The aim is to develop problem-solving skills, attitudes and effective communication.

There will be occasions when pupils are struggling to remember, for example, the name of a shape or the counting sequence. Our advice in these situations is to facilitate the problem-solving aspect. Please do help the pupils to stay on task with the problem.

Let's Go – enabling prompt and extension activity

> **Let's go!**
>
> Patterned paper plates (have prepared a number of plates similar to the ones in the picture).
>
> Our plates have been cut up by mistake and we have a jumble of pieces. **I wonder what we can do? What do you think we need to do to get started? Can you estimate how many whole plates we would be able to make? What makes you think that?**
>
> ### Enabling prompt
>
> Provide two 'chopped up' paper plates (with two different patterns) per pair of pupils. Allow time for the pupils to talk about what they see and what they might do. Let them explore and investigate at their own pace. **What shapes can you see? How many different patterns can you see? Can you describe the patterns? Do you think they are repeating patterns?**
>
> ### Extension activity
>
> Increase the number of 'chopped up' paper plates per pair of pupils or add complexity to this task by either
>
> * using similar plate patterns or
> * having a plate with five pieces
>
> Allow time for the pupils to talk about what they see and what they might do.

The **Let's go** section contains the problem to be solved. It is important that pupils are allowed to puzzle initially with minimal, and on occasion no, input from the teacher. This will support deeper discussion and support a focus on metacognition as pupils progress through the problem. Some problems are open ended in nature and have multiple solutions; others have a definite answer.

Creativity is at the heart of problem solving. Even problems with a clear solution provide opportunities for creativity. Pupils should be encouraged to play, experiment, explore and question. The educator's role is to observe closely, in order to understand the pupil's train of thought, and ask questions which develop and extend their ideas.

Some pupils may struggle with the problem. For these pupils, we have created **Enabling prompts**. Enabling prompts may include questions to support their initial ideas, for example **Tell me what you are thinking,**

What can you see? An **Enabling prompt** may also suggest ways to simplify the original problem. It may be that the pupil does not have the mathematical knowledge required for the problem, or perhaps they lack resilience or confidence to try out something new. It is important that the pupil is allowed to engage with the problem at their own 'level' and is not left to drift. Providing a simpler example, help from a partner or support from a member of staff can make the problem accessible to these pupils and ensure they feel included.

Some pupils will solve the problem quickly, demonstrating mastery of the task. The **Extension activities** can be used to stretch these pupils. 'What if….' type questions should be used to extend these pupils' thinking and deepen their understanding.

Let's Discuss

> ### Let's discuss!
>
> Amman and Isla had been practising rolling two 1–6 dice. They used adding and taking away to solve the problem. They looked at the first number at the base of the mountain and decided that they couldn't make one by adding together any of the dice numbers – the closest they could get was two by adding one and one. Isla said she thought there was more than one way to make one by taking away. She said 2 – 1 = 1 and 3 – 2 = 1. **Do you think there are other solutions that they haven't mentioned?** Amman and Isla wondered if this was the only number on the mountain that had no adding solution. **What do you think?**
>
> Finlay and Nuria found that some numbers only had one solution. They thought that 12 was the only number to have one adding solution (6 + 6). **Do you think they are correct?** They wondered if odd numbers had more solutions than even ones. **Do you think that is true?**

In the **Let's discuss** section we have provided a narrative on some of the possible strategies that pupils may use to solve the problem. We have imagined how the Leckie children: Amman, Findlay, Isla and Nuria, might have solved the problem (or even partially solved the problem). These can be used in a number of ways; to provide prompts to help the pupils, to pose questions to the pupils to deepen their understanding or to help you recognise when pupils are either on the right track or searching down a blind alley. Please note that this is not an exhaustive list of expected strategies, nor a directive that pupils must follow these routes to the solution. In problem solving there are many, many strategies that can be used. What we have found is that the pupils will develop a range of problem-solving skills that they will come to rely on. The challenge is to support them to grow these strategies and to see if they can apply them to new and unfamiliar contexts.

Let's Check

> ### Let's check!
>
> The solution will be to locate and place the appropriate pieces for each plate, while matching the pattern. The plates should be re-assembled to make a circle, the patterns should match and there should be no gaps. These questions can be used with the pupils to explore their knowledge, strategies and understanding of the problem:
>
> - **Can you see a pattern? What do you notice? Can you see anything that is happening again and again? What shapes can you see? What shape is the plate? Can you recognise the shapes or markings on the plates?**

The **Let's check** section provides suggested solutions to the **Let's go** problem, the **Enabling prompt** and the **Extension activity**, except for situations where the problem is more open ended and there are multiple solutions. In these cases, we have tried to give an example of the possible solutions to look out for. We advocate that the pupils have an opportunity to share their solutions with their peers before the teacher confirms if an answer is correct.

Let's Reflect

Let's reflect

This challenge allows the pupils to practise adding and subtracting in a new and unfamiliar context. Encourage them to follow their own lines of enquiry, discuss their ideas and ask questions. Questions or musings are important because they can lead to new lines of enquiry that pupils are interested to investigate. This activity is extended in the next Wonder Jar Challenge.

To promote metacognition, the **Let's reflect** section provides specific guidance for each problem task. As pupils become more experienced with solving problems, they can be prompted to think back to similar situations, to see if they can remember and apply a strategy that they have used successfully in the past.

Teacher Guide Reference

Teacher Guide Reference

- Chapter 2 – Number and Number Processes – Number Recognition and Place Value
- Chapter 3 – Number and Number Processes – Four Operations

Whilst solving the problem you may notice gaps in the pupil's mathematical knowledge or skills. These links will help you to plan revision for these missing skills.

Planning and Carrying out the Problem Solving Activity

The following is a series of planning prompts, based on our experience of facilitating problem solving with our own pupils.

Prior to commencing, check that the pupils have the mathematical knowledge necessary to access the problem and that they understand the language used. Ensure that any physical resources the pupils might need are readily available.

Some problems have a real-life context, others are fantasy contexts. Several centre around outdoor spaces but can be adapted for indoor settings, if necessary. The problem may present a situation that pupils are familiar with, or an unfamiliar context to pique their interest. You might want to introduce new contexts to the pupils in the days running up to the problem-solving activity. If you feel that changing the context would help engage your pupils, then do so. Allowing for personalisation and choice is integral to the problem-solving process. Many pupils will see the activity as play and we encourage this view of mathematics as an enjoyable and creative experience.

After you introduce the problem to the pupils your role is one of facilitator. Observe pupils' closely as they engage with the problem. How do they tackle it? What does this tell you? Be ready with the questions you will use as you follow and listen to the children talking, playing and acting out the problem. For pupils who are struggling you can use the **Enabling prompts** to help them progress with the problem-solving activity.

Here are some suggested questions that you can use to support your pupils:

What do you need to find out?	Is there anything you can't change? Why is this?
How are you going to start?	Why do you think that?
How are you going to solve the problem?	Is that always true? Tell me why you think that.
What do you know already that can help?	How do you know ..?
What do you notice?	What else could you do?
What do you wonder?	Why did you choose…?
What have you found out so far?	Let's try another way!
Can you show me a different way?	Have you seen something like this before?
Do you need anything else to help you?	What is the same/different?
Can you tell me more about ..?	What would help you…?
What would happen if ..?	Why didn't that work? Is there another way?
What can you change?	

As the problem-solving activity comes to an end, bring the children together to talk about how they solved the problem, whether or not they had done something similar in the past and what new skills or knowledge they have developed.

References

Education Scotland (2017) *Benchmarks Numeracy and Mathematics* available at https://education.gov.scot/nih/Documents/NumeracyandMathematicsBenchmarks.pdf

Polya, G (1945) *How to Solve it*. Princeton University Press

Scottish Executive (2006) *Building the Curriculum 1: The Contribution of Curricular Areas*. Edinburgh: Scottish Executive. https://education.gov.scot/Documents/btc1.pdf

Scottish Government (2011) *Mathematics Principles and Practice*. Edinburgh: Scottish Government. https://education.gov.scot/Documents/mathematics-pp.pdf

Wonder Jar 1 – The Paper Puzzle (Monday)

Experiences and Outcomes

- *MNU 0-01a* – *I am developing a sense of size and amount by observing, exploring, using and communicating with others about things in the world around me.*
- *MNU 0-02a* – *I have explored numbers, understanding that they represent quantities, and I can use them to count, create sequences and describe order.*

Numeracy and Mathematical Skills

- **Uses mathematical vocabulary and notation** – uses developmentally appropriate mathematical vocabulary
- **Mental agility** – manipulates numbers
- **Determine the reasonableness of a solution** – routinely uses estimation skills

Resources

- A wonder jar
- Folded pieces of paper (number of pieces will depend on the age and stage of pupils – this example shows 12)
- Access to materials to help check the count, e.g. ten frames, bead strings, chalk, counters, etc.

Before they start...

The pupils should have:

- experience of estimating in a range of scenarios, e.g. using loose parts, patterned wrapping paper, etc.
- experience of organising and counting collections of objects

Let's go!

Show the pupils the wonder jar and ask **I wonder how many pieces of paper are in the jar?**

Use these questions to prompt the pupils' thinking: **I wonder if you could estimate how many pieces of paper are in the jar? Do you think it might be more than 5/less than 10? Another number?**

Once the pupils have all had a chance to give an estimate, ask them: **How might we find out how many pieces of paper are in the jar? What would be the best way to check and see what the total number of pieces of paper actually is?**

Before emptying the jar, take a photograph to use as a comparator in the next lesson.

Enabling prompt

Use different coloured pieces of paper in the jar, or a smaller number of pieces, as this makes it easier for the pupils to gauge 'how many'. Encourage the use of empty ten frames to assist with organising the count.

Extension activity

To provide more challenge, use one colour of paper, increase the number of pieces or have different sizes of paper in the jar.

Let's check!

The pupils should be able to make a reasonable estimate (within an acceptable range) and then suggest an effective way to find the actual amount.

Let's discuss!

Isla estimated less than 10 but more than 5. She suggested that the paper should be tipped out of the jar and counted into a bowl.

Finlay estimated more than 10; he said that the pieces of paper could be taken from the jar and placed on to ten frames – one piece of paper for each square on the ten frame. He took out two ten frames because he thought there were less than 20 but more than 10.

Nuria estimated 9 pieces of paper. She decided to place each piece of paper onto a number track to help her see how many there actually were.

Amman estimated 20 pieces of paper; he said that for every piece of paper taken out of the jar, a bead could be moved along the bead string.

What was your estimate? Was your estimate close to that of Finlay, Nuria, Amman or Isla? Do you think you would use any of the models suggested to show your counting? Is there a better way to check the count?

Let's reflect!

Pupils who are struggling to provide a reasonable estimate should be given opportunities to compare collections visually, by grouping and counting or lining up and matching. The solutions above show that the pupils can give a reasonable estimate and are able to select and use a range of manipulatives; however, not all of the manipulatives will be as efficient if the number range increases. Work with pupils to explore the most effective manipulatives to tackle the problem.

Teacher Guide Reference

- Chapter 1 – Estimation and Rounding
- Chapter 2 – Number and Number Processes – Number Recognition and Place Value

Wonder Jar 2 – Size Matters (Tuesday)

Experiences and Outcomes

- *MNU 0-01a – I am developing a sense of size and amount by observing, exploring, using and communicating with others about things in the world around me.*
- *MNU 0-02a – I have explored numbers, understanding that they represent quantities, and I can use them to count, create sequences and describe order.*

Numeracy and Mathematical Skills

- **Uses mathematical vocabulary and notation** – uses developmentally appropriate mathematical vocabulary
- **Mental agility** – manipulates numbers
- **Determine the reasonableness of a solution** – routinely uses estimation skills

Resources

- A wonder jar
- Pieces of macaroni or other small pasta, small pebbles or beads (this example shows 19 pieces of macaroni)
- Ten frames, chalk, bead strings, etc.
- Picture of Monday's Paper Puzzle (for comparison)

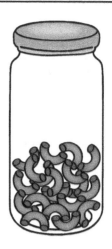

Before they start...

The pupils should have:

- experience of estimating in a range of scenarios (including measurement)
- tried out The Paper Puzzle wonder jar problem (Monday)
- experience of organising and counting collections of objects
- an understanding of more than / less than

Let's go!

Ask the pupils how many pieces of macaroni they think might be in the jar. **Do you think there are more pieces of macaroni today than there were pieces of paper yesterday?** Let's look at a picture of yesterday's wonder jar. **Do you think there are more or less? Can you tell me why you think that?**

Enabling prompt

This is a direct comparison between items in a photograph or image and concrete objects in a jar. Remind the pupils of the total from Wonder Jar 1 – The Paper Puzzle. Provide opportunities for visual comparison. Encourage the pupils to explore collections, e.g. of up to 10 small objects (buttons, beads, toy animals, pom-poms) and compare them visually. **What do you notice?** Encourage the use of comparative language: more than, less than, the same.

Extension activity

To create more of a challenge, have a **selection** of small items in the wonder jar such as a mixture of small pasta, small pebbles and small buttons.

Let's check!

There is more macaroni than pieces of paper, but when comparing two collections of objects that are dissimilar in size or shape, it is not uncommon for some pupils to equate the fullness of the jar with there being more.

Let's discuss!

Isla estimated that there was less macaroni than pieces of paper because there was more space left in the jar. **Do you agree with Isla? Why do you think that?** She didn't suggest tipping out the things in the jar today as she thought the pieces would roll onto the floor. **Can you suggest a way to help Isla count the macaroni?**

Finlay thought that there was the same number of pieces of macaroni as paper. He suggested placing the pasta onto ten frames. He said it had worked for him on Monday and it was easy to count. **Is Finlay correct? Why might he find it easier to count using ten frames?**

Nuria thought there was less macaroni than paper. She counted the pasta into a small tub to check. Each time she put a piece of pasta in the pot she marked a 1 on her whiteboard. **Do you think Nuria chose a good way to keep track of her count? Why do you think that?**

Amman thought there was much less in the jar today. He thought that the macaroni was about half the number of pieces of paper. Amman used his bead string as he found it was a useful way to check his estimate. **Did you use a bead string like Amman? How did you count the pieces of macaroni? Was your estimate a good estimate? What makes you say that?**

Let's reflect!

All the children thought that there was less macaroni than paper. It challenged their thinking when they counted and found that there was actually more in the jar. They were unable to state which collection had more when there was a significant visual difference due to the size of the objects in the jar. Provide more opportunities similar to Wonder Jar 1 (Monday), where the pupils can explore more or less in terms of capacity. Ensure that the pupils can 'see' the difference in amount using the same objects in each jar before taking the next step and comparing different sized objects. Help the pupils to compare visually by using matching or lining-up strategies.

Teacher Guide Reference

- Chapter 1 – Estimation and Rounding
- Chapter 2 – Number and Number Processes – Number Recognition and Place Value

Wonder Jar 3 – Ribbon Jumble (Wednesday)

Experiences and Outcomes

- *MNU 0-11a* – *I have experimented with everyday items as units of measure to investigate and compare sizes and amounts in my environment, sharing my findings with others.*
- *MNU 0-20b* – *I can match objects and sort using my own and others' criteria, sharing my ideas with others.*

Numeracy and Mathematical Skills

- **Justify choice of strategy used** – shows and talks through their thinking
- **Uses mathematical vocabulary and notation** – uses developmentally appropriate mathematical vocabulary

Resources

- A wonder jar
- A selection of ribbons of different lengths, patterns and colours. Include a similar object such as a coloured lace or piece of string

Before they start...

The pupils should have experience of:

- sorting objects using their own and others' criteria

Let's go!

Wednesday.

In the wonder jar today, we have a jumble of ribbons. We need to sort them. **Do the ribbons all look the same? What makes them the same? How are they different? How will you sort the ribbons? How many sets do you think you can make?** Questions alluding to less obvious attributes may also help pupils formulate ideas about sorting, e.g. **Are all the ribbons shiny? Do you think some ribbons might be rough/smooth?**

Enabling prompt

Encourage the pupils to remove the ribbons from the jar. Talk about what they can see. If necessary, prompt the pupils to look for similarities and differences in colour, pattern and length. Ask – **Can you sort the ribbons into two bundles? How did you sort them?**

Extension activity

Sort the ribbons and then re-sort them using a different criterion, e.g. If the pupils initially sorted the ribbons using colour (red and not red), they could then re-sort using pattern as their criterion (checked pattern / spotty pattern / plain). Challenge the most able to sort the ribbons using two criteria.

Let's check!

Tip the ribbons out of the jar.

There are multiple ways that the pupils could sort them, e.g. patterned and not patterned, ribbon and not a ribbon, spots and no spots. The children may also choose to sort by length or width – wide and narrow or long and short. Observe how the pupils compare the ribbons and listen for correct use of mathematical vocabulary.

Let's discuss!

Finlay decided that he would sort the ribbons into two groups – ribbon and not a ribbon. He set the one yellow lace to the side and put the ribbons in a bundle. He noticed that the lace was not as wide as the ribbons. **Did Finlay manage to sort the collection? Is this what you did? Did you do something different?**

Nuria sorted the ribbons into two bundles – those that she liked and those that she didn't like. **Do you think that is a good way to sort the ribbons?**

Amman decided he would sort the ribbons by looking at the patterns. He decided to use patterns that had squares and patterns that did not have squares. Once he had taken a picture of his sorting, he re-sorted them by colour. He decided to have ribbons that had red and ribbons that had no red on them. **Did you sort your ribbons in more than one way? What did you look for?**

Isla decided to look at the length of the ribbons. She sorted the bundle into long and short ribbons. Some ribbons were easy to sort but some needed her to compare their lengths. She carefully lined the ribbons up to check. Isla noticed when she was comparing the lengths of the ribbons that two were exactly the same length. She decided she would re-sort by having two ribbons that were the same length and a big group of ribbons that were different lengths. **Do you think Isla's idea was a good one? How did you sort the ribbons?**

Let's reflect!

Vocabulary plays an important part in sorting. Model appropriate mathematical vocabulary at all times and encourage the pupils to use it during their exploration and play, and when sharing their ideas. Maths vocabulary can help pupils make links across different areas of maths, across the curriculum and make connections with real-life.

Sorting is an important mathematical skill in the early stages of data handling. Pupils need to have a sound understanding of the attributes they are using for comparison. Sorting is different to matching as it involves re-organising the whole collection into subsets. Finlay, Amman and Isla were all able to describe the attribute they were sorting by, whereas Nuria found it difficult to describe why she sorted the ribbons the way she did. Next steps would be to encourage the pupils to find a different way to sort the ribbons or to try to identify someone else's sorting criteria.

Teacher Guide Reference

- Chapter 7 – Measurement
- Chapter 11 – Data and Analysis

Wonder Jar 4 – What Numbers Were Rolled (Thursday)

Experiences and Outcomes

- *MNU 0-02a – I have explored numbers, understand that they represent quantities, and I can use them to count, create sequences and describe order.*
- *MNU 0-03a – I use practical materials and can 'count on and back' to help me understand addition and subtraction, recording my ideas and solutions in different ways.*

Numeracy and Mathematical Skills

- **Selects and communicate processes and solutions** – shares thinking; verbalises or demonstrates thought processes
- **Justify choice of strategy used** – shows and talks through their thinking
- **Mental agility** – knowledge of number facts

Resources

- 2 × 1–6 spot dice
- 2 × 1–6 numbered dice (Extension activity)
- Resource sheet - Wonder Jar 4

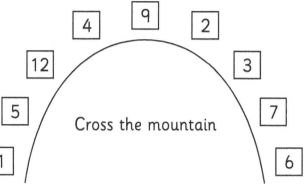

Cross the mountain

Choose appropriate numbers for age and stage of pupils

Before they start...

The pupils should be able to add within 12 and subtract within 6.

The pupils should be able to identify and recognise numbers up to 12

Let's go!

Have a rolled-up version of the problem below in the wonder jar. Also display the diagram on the interactive white board (IWB) for the pupils to refer to.

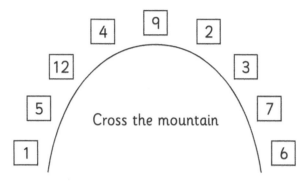

Tell the pupils that they have to make their way across the mountain. In order to move up, over and down the mountain they have to think of ways to make the numbers in the boxes by adding or subtracting the numbers displayed after rolling two 1–6 spot dice. **If we have two dice, what numbers could we roll to give the answer 6? Did you add or did you subtract the numbers on the dice? If we have two dice, what numbers could we roll to give the answer 1? Did you add or did you subtract the numbers on the dice? Can you find more than one way to cross the mountain?**

Share solutions as a class.

Enabling prompt

Before attempting the problem, provide two dice for the pupils to practise rolling and recording the scores arrived at when they add the two numbers together or subtract one number from the other. Pupils who are unable to subitise should be allowed to touch count the dots. Accept one solution per number.

Extension activity

Encourage the pupils to find as many solutions as they can for each boxed number (both addition and subtraction facts). **Can you note down all the solutions you can think of before moving to the next number?** Increase the challenge further by giving the pupils numbered, rather than spotted, dice.

Let's check!

There are a limited number of solutions for each boxed number and being able to produce the solution relies on the pupils having a good working knowledge of number bonds. This challenge provides:

- a contextualised opportunity to explore the commutative property, e.g. $4 + 1 = 1 + 4$
- an opportunity to investigate that for some numbers there are lots of ways of making the answer, e.g. 7

Let's discuss!

Amman and Isla had been practising rolling two 1–6 dice. They used adding and taking away to solve the problem. They looked at the first number at the base of the mountain and decided that they couldn't make one by adding together any of the dice numbers – the closest they could get was two by adding one and one. Isla said she thought there was more than one way to make one by taking away. She said 2 – 1 = 1 and 3 – 2 = 1. **Do you think there are other solutions that they haven't mentioned?** Amman and Isla wondered if this was the only number on the mountain that had no adding solution. **What do you think?**

Finlay and Nuria found that some numbers only had one solution. They thought that 12 was the only number to have one adding solution (6 + 6). **Do you think they are correct?** They wondered if odd numbers had more solutions than even ones. **Do you think that is true?**

Let's reflect!

This challenge allows the pupils to practise adding and subtracting in a new and unfamiliar context. Encourage them to follow their own lines of enquiry, discuss their ideas and ask questions. Questions or musings are important because they can lead to new lines of enquiry that pupils are interested to investigate. This activity is extended in the next Wonder Jar Challenge.

Teacher Guide Reference

- Chapter 2 – Number and Number Processes – Number Recognition and Place Value
- Chapter 3 – Number and Number Processes – Four Operations

Wonder Jar 5 – Rolling Dice (Friday)

Experiences and Outcomes

- *MNU 0-02a – I have explored numbers, understand that they represent quantities, and I can use them to count, create sequences and describe order.*
- *MNU 0-03a – I use practical materials and can 'count on and back' to help me understand addition and subtraction, recording my ideas and solutions in different ways.*

Numeracy and Mathematical Skills

- **Select and communicate processes and solutions** – shares thinking; verbalises or demonstrates thought processes
- **Justify choice of strategy used** – shows and talks through their thinking
- **Mental agility** – knowledge of number facts
- **Reason algebraically** – finds the unknown quantity

Resources

- Dice to explore (optional)

- Resource sheet – Wonder Jar 5

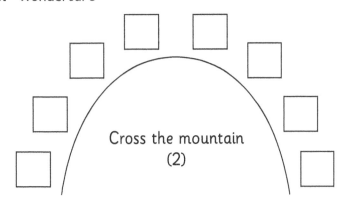

Cross the mountain
(2)

Before they start...

The pupils should be able to:

- add two numbers totalling up to 12 and subtract within 6
- identify and recognise numbers within 12

For the Enabling prompt:

- add two numbers totalling up to 6 and subtract within 3 (use dice with 2 × 1, 2 × 2 and 2 × 3)

For the Extension activity:

- add two numbers totalling up to 15/18 and subtract within 9/12 (use a 1–12 dice and a 1–6 dice)

Let's go!

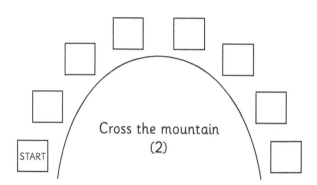

Cross the mountain (2)

START

Cross the Mountain

Take the rolled-up challenge out of the wonder jar. Show the children the 'empty' Cross the Mountain diagram and ascertain that they remember tackling Thursday's wonder jar problem. Tell them their challenge today is to create their own problem, by thinking of numbers to place in the empty boxes of Cross the Mountain.

Ask: **What is the biggest number you can make with two 1–6 dice? What is the smallest number you can make with two 1–6 dice? What might be the biggest number you could write on your empty boxes? What might be the smallest number you can write on your empty boxes? Are there any numbers that you can't make? Why is that?** Challenge pupils to work in pairs to find all possible solutions. They should complete the mountain, note their solutions on a separate piece of paper then swap mountains with another pair. Can they check the other pair's mountain to see if all the solutions were found? Display the problem on the IWB and encourage pairs of pupils to discuss and share their thinking.

Enabling prompt

Provide pupils with dice numbered 1–6 and numbered 1–3. **What is the biggest number that can be made by adding together the score from each of the dice? What is the smallest number that can be made by subtracting the scores from each of the dice?** Some pupils many require spot dice to be provided to help with counting. The pupils may find it useful to draw dot patterns in the empty boxes to help with the partitioning of numbers.

Extension activity

Tell the pupils that they can choose one 1–6 dice and either of a 1–9 or 1–12 dice. Can they think of the biggest and smallest number that could be made when using a 1–6 and either a 1–9 or 1–12 dice? Consider adding extra boxes on the mountain to fill. Can they make their own Cross the Mountain challenge for others to try?

Let's check!

Solutions will vary depending on the type of dice used.

Possible solutions:

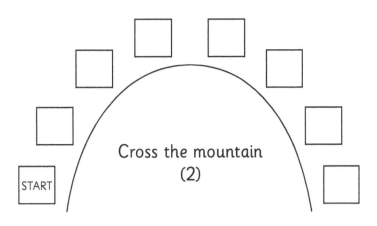

Cross the mountain (2)

START

Using 1–6 and 1–9 Dice

15 = 9+6 (6+9)

14 = 9+5 (5+9), 8+6 (6+8)

13 = 9+4 (4+9), 8+5 (5+8), 7+6 (6+7)

12 = 9+3 (3+9), 8+4 (4+8), 7+5 (5+7), 6+6

11 = 9+2 (2+9), 8+3 (3+8), 7+4 (4+7), 6+5 (5+6)

10 = 9+1 (1+9), 8+2 (2+8), 7+3 (3+7), 6+4 (4+6), 5+5

9 = 8+1 (1+8), 7+2 (2+7), 6+3 (3+6), 5+4 (4+5)

8 = 7+1 (1+7), 6+2 (2+6), 5+3 (3+5), 4+4, **9–1**

7 = 6+1 (1+6), 5+2 (2+5), 4+3 (3+4), **9–2, 8–1**

6 = 5+1 (1+5), 4+2 (2+4), 3+3, **9–3, 8–2, 7–1**

5 = 4+1 (1+4), 3+2 (2+3), **9–4, 8–3, 7–2, 6–1**

4 = 3+1 (1+3), 2+2, **9–5, 8–4, 7–3, 6–2, 5–1**

3 = 2+1 (1+2), **9–6, 8–5, 7–4, 6–3, 5–2, 4–1**

2 = 1+1, **8–6, 7–5, 6–4, 5–3, 4–2, 3–1**

1 = **7–6, 6–5, 5–4, 4–3, 3–2, 2–1**

0 = **6–6, 5–5, 4–4, 3–3, 2–2, 1–1**

Using 1–6 and 1–12 Dice

18 = 12+6 (6+12)

17 = 12+5 (5+12), 11+6 (6+11)

16 = 12+4 (4+12), 11+5 (5+11), 10+6 (6+10)

15 = 12+3 (3+12), 11+4 (4+11), 10+5 (5+10), 9+6 (6+9)

14 = 12+2 (2+12), 11+3 (3+11), 10+4 (4+10), 9+5 (5+9), 8+6 (6+8)

13 = 12+1 (1+12), 11+2 (2+11), 10+3 (3+10), 9+4 (4+9), 8+5 (5+8), 7+6 (6+7)

12 = 11+1 (1+11), 10+2 (2+10), 9+3 (3+9), 8+4 (4+8), 7+5 (5+7), 6+6

11 = 10+1 (1+10), 9+2 (2+9), 8+3 (3+8), 7+4 (4+7), 6+5 (5+6), **12–1**

10 = 9+1 (1+9), 8+2 (2+8), 7+3 (3+7), 6+4 (4+6), 5+5, **12–2, 11–1**

9 = 8+1 (1+8), 7+2 (2+7), 6+3 (3+6), 5+4 (4+5), **12–3, 11–2, 10–1**

8 = 7+1 (1+7), 6+2 (2+6), 5+3 (3+5), 4+4, **12–4, 11–3, 10–2, 9–1**

7 = 6+1 (1+6), 5+2 (2+5), 4+3 (3+4), **12–5, 11–4, 10–3, 9–2, 8–1**

6 = 5+1 (1+5), 4+2 (2+4), 3+3, **12–6, 11–5, 10–4, 9–3, 8–2, 7–1**

5 = 4+1 (1+4), 3+2 (2+3), **11–6, 10–5, 9–4, 8–3, 7–2, 6–1**

4 = 3+1 (1+3), 2+2, **10–6, 9–5, 8–4, 7–3, 6–2, 5–1**

3 = 2+1 (1+2), **9–6, 8–5, 7–4, 6–3, 5–2, 4–1**

2 = 1+1, **8–6, 7–5, 6–4, 5–3, 4–2, 3–1**

1 = **7–6, 6–5, 5–4, 4–3, 3–2, 2–1**

0 = **6–6, 5–5, 4–4, 3–3, 2–2, 1–1**

Let's discuss!

Amman and Finlay chose to use 1–6 and 1–3 dice. They thought about the biggest and smallest numbers they could make; Amman said the biggest number they could make was 9. **Do you think Amman was correct? What numbers could have been on the dice?** Finlay wondered if the smallest number was 3 but then he wondered if they could put zero in one of the boxes. **Do you agree with Finlay? Can you tell me why?**

Isla and Nuria looked at 1–6 and 1–9 dice. They wanted to put numbers in the boxes so that each box would have both adding and subtracting solutions. They decided that 2 was a good number. **Can you help them think of more numbers that they could have used?**

Let's reflect!

The pupils are being given an opportunity to think more closely about addition and subtraction in a context. This activity is a good assessment challenge – look carefully for pupils who are still counting from one rather than counting on or using known facts, or those who are struggling to count back from, or count down to, a number. This may indicate the need for activities that involve:

- screening items
- practice in reciting backwards number word sequences

Teacher Guide Reference

- Chapter 2 – Number and Number Processes – Number Recognition and Place Value
- Chapter 3 – Number and Number Processes – Four Operations

Balloons

Experiences and Outcomes

- *MNU 0-02a* – *I have explored numbers, understand that they represent quantities, and I can use them to count, create sequences and describe order.*
- *MNU 0-03a* – *I use practical materials and can 'count on and back' to help me understand addition and subtraction, recording my ideas and solutions in different ways.*

Numeracy and Mathematical Skills

- **Interpret questions** – selects the relevant information; draws diagrams
- **Select and communicate processes and solutions** – verbalises or demonstrates thought processes
- **Justify choice of strategy used** – shows and talks through their thinking
- **Use mathematical vocabulary and notation** – uses developmentally appropriate mathematical vocabulary
- **Mental agility** – knowledge of number facts; manipulates numbers

Resources

- Resource sheet – Balloons (optional)

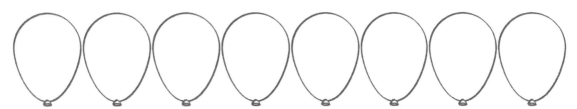

- Paper and colouring pencils / crayons
- Two colours of counters (three colours for Extension activity)

Before they start...

Pupils should have experience of exploring numbers in a wide variety of contexts (counting principles) and be able to:

- partition (within 8)
- subitise regular and irregular dot patterns (up to 8)

Let's go!

Nuria has 8 balloons; some of Nuria's balloons are red and some of her balloons are green. **I wonder how many red balloons Nuria has, and how many green balloons she has? Draw a picture of Nuria's balloons to show me what you think. Can you find another way to colour the balloons? How many different ways can you find?**

For pupils unable to draw balloons, provide a template sheet or use coloured counters to represent the balloon combinations.

Enabling prompt

To help pupils focus on different ways to make the same amount:

- Have prepared a balloon template showing four green and four red balloons. **How many green balloons can you see? How many red? How many balloons altogether? What if there was one red balloon, how many green balloons would there be? What other ways can you find?**

OR

- Provide opportunities for them to investigate partitioning collections of counters, cube towers, etc. For example, ask: **Build a tower of five cubes and split it into two smaller towers. How many cubes do you have in each tower?**

Extension activity

Add some challenge by increasing the number of balloons and/or introduce another colour, e.g. Amman has eight balloons. He has three different colours of balloons – red, blue and green. **How many different colour combinations can you make? Draw or use counters to show the ways you have found.**

Let's check!

Solutions for two colours:

Green Balloons	Red Balloons
7	1
6	2
5	3
4	4
3	5
2	6
1	7

Let's discuss!

Finlay took some red and some green counters. He thought he would make a pattern of red, green, red, green and stopped when he had eight counters altogether. He found that this gave him four green and four red. He then tried to make a red, red, green, green pattern with his eight counters and was surprised to find this gave him the same answer of four green and four red. He explored other patterns and found red, green, green, green gave him two reds and six greens. **Can you think of other patterns that might give different answers?**

Isla decided that she might be able to use stories of eight to help her. She knew that Nuria had eight balloons that were a mixture of red balloons and green balloons. She drew eight oval shapes to represent the balloons. She thought of different ways to make eight: 7 and 1; 6 and 2; 5 and 3; and 4 and 4. **Do you think Isla has found all the different ways?** She coloured her ovals – seven red and one green. Then she drew another eight ovals and coloured six red and two green, etc. Isla saw a pattern as she coloured. **How many different ways do you think there are to colour the balloons? Can you explain your thinking? Did you see a pattern, like Isla did?**

Let's reflect!

Come together as a group or class once the pupils have had time to illustrate their thinking. **How did the pupils choose to solve the problem – which strategy did they start with? Did any of the pupils start by using counters, did they spot a number pattern or did they notice the partitioning patterns? Let's look and see how many different combinations of red balloons and green balloons we have found. Do you think we have found all the possible solutions?**

Although some manipulatives are best suited for a particular problem, others are versatile and easily adapted. Pupils should have experience of using a range of manipulatives to solve different problems. **Ask the pupils if they can think of other resources that could be used to solve this problem.**

Teacher Guide Reference

- Chapter 2 – Number and Number Processes – Number Recognition and Place Value
- Chapter 3 – Number and Number Processes – Four Operations

Bean Bag Challenge

Experiences and Outcomes

- *MNU 0-03a – I can use practical materials and can 'count on and back' to help me understand addition and subtraction, recording my ideas and solutions in different ways.*
- *MNU 0-07a – I can share out a group of items by making smaller groups and can split a whole object into smaller parts.*

Numeracy and Mathematical skills

- **Interpret questions** – selects the relevant information, interprets data, draws diagrams
- **Select and communicate processes and solutions** – shares thinking; verbalises and demonstrates thought processes
- **Justify choice of strategy used** – shows and talks through their thinking
- **Links mathematical concepts** – transfers learning from one area to another
- **Uses mathematical vocabulary and notation** – uses developmentally appropriate mathematical vocabulary

Resources

- Selection of concrete materials, including counters and empty ten frames
- Selection of number tiles and number tracks
- Resource sheet – Bean Bag Challenge: blank templates (enlarged for enabling prompt)

Before they start...

Pupils should have experience of:

- playing games such as this in PE or in the outdoor area (see Bean Bag Game (Task 4 - Playing Games - Primary Maths for Scotland Assessment Book))
- addition bonds to 6.

Let's go!

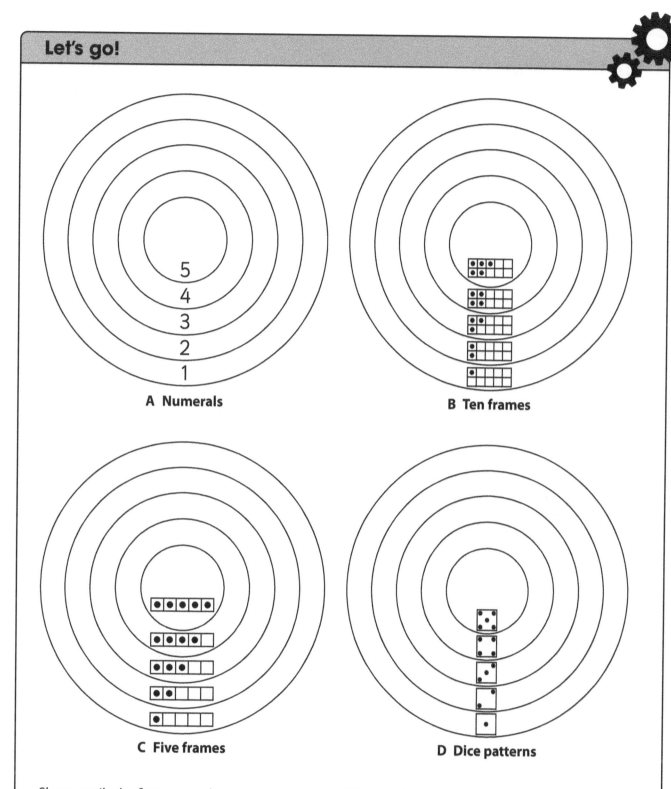

A Numerals

B Ten frames

C Five frames

D Dice patterns

Show pupils the first numerals target board image. Tell them that one of Amman's friends was playing a throwing game in the playground using two bean bags and a target. He scored 6 using his two bean bags. I wonder where his bean bags landed on the target to give him a score of 6. Brainstorm with the pupils. **What do we want to find out? What do you think we should do first to solve the problem? Do we need to use anything to help us? How many different ways can you find to make 6 using the numbers on the target board?**

Enabling prompt

Provide pupils with concrete materials (see Resources) and encourage them to investigate ways of making 6. Alternatively, templates that are slightly larger could be provided and pupils encouraged to draw the bean bags in the appropriate rings of the target (pupils unable to draw bean bags could be given, for example a bingo stamper to make a representative shape). Pupils who are still struggling could be given target boards B, C and D.

Extension activity

For the extension task, Amman's friend has been given three bean bags. **Where would his bean bags have landed to make a score of 6?**

Provide pupils with

- a picture (IWB) of the challenge
- access to concrete manipulatives
- wipe-clean marker and whiteboard
- templates to record their solutions on

Encourage the pupils to explore the task, find some solutions and record their findings. Vary the task by changing the target number to a number greater than 6 and/or challenge the pupils to work out the largest score possible with two or three bean bags.

Let's check!

A number of ways of finding solutions are possible, depending on what the pupils choose to use. The approaches the pupils use, and what they say, will highlight their knowledge of addition bonds within 6. Discussion of solutions should highlight the commutative nature of addition and doubles, e.g. 1 + 5, 5 + 1; 2 + 4, 4 + 2; 3 + 3. For the extension task, the pupils will be exploring adding three numbers to make 6. Solutions should include those involving a 0 to represent one bean bag missing the target, e.g. 5 + 1 + 0, 3 + 3 + 0, 4 + 2 + 0. Other possible solutions may include, e.g. 4 + 1 + 1, 3 + 2 + 1, 2 + 2 + 2.

Let's discuss!

Here are some solutions found by Nuria, Finlay, Amman and Isla. I wonder if anyone solved the problem the same way as they did.

Enabling

Nuria took six counters and laid them in a line. She split her counters to show what she knew about addition: six counters and zero counters, five counters and one counter, etc. **What ways can you think of to make six? Did you use counters like Nuria?**

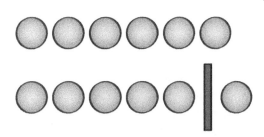

Did you start the way Nuria did? How many solutions did you find?

Nuria drew bean bags onto the target template but was confused when she couldn't find 6 on the target. **Would having a six on the target help?**

Amman counted out six cubes and built them into a tower. He split his tower into 5 and 1, then 4 and 2, etc. As he found his solution, he drew bean bags on the target template. He completed three templates. **Did you use cubes like Amman? Did you build them into a tower? Do you think that was a good way of finding a solution? Do you think Amman found all the solutions?** He asked if the bean bags were different colours – he said if they were then he would maybe have more answers. **Do you think if there were two different colours of bean bags, it would give more solutions?**

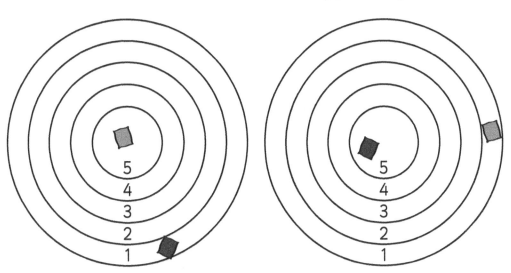

Extending

Finlay thought about ways of making 6 using three numbers; he chose to use counters to help him solve the problem. **What did you choose to use? What did you do first?** He looked at the target to check what numbers he could use – he could use 5, 4, 3, 2 and 1. He drew a box on his whiteboard and used this to help him keep track as he shared his six counters into three different groups.

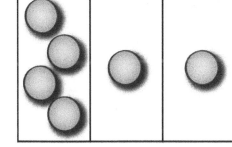

Did anyone solve the problem like Finlay? Tell us what you did. What did you find? Finlay recorded his findings on the target template. He knew some facts for adding two numbers but had to count on to add the third number. **Were you able to make six using three numbers? How did you do it? What helped you keep track of what you were doing?**

Isla decided to write out the ways she knew of making 6 using three numbers. She started writing $6 + 0 + 0$, $5 + 1 + 0$ but then realised there was no 6 on the target and that she had to use 3 bean bags so she decided to start again. This time she started with $4 + 1 + 1$ and then $3 + 2 + 1$, etc. **Did you use number facts to solve the problem?** Isla checked her calculations on a number line. She drew the solutions onto the template. Isla then thought about when she had played a target game outside. She had missed the target a few times when she was throwing the bean bags and one landed outside the rings. This made Isla wonder if she had all the possible solutions. She got more target templates and drew a bean bag outside the ring.

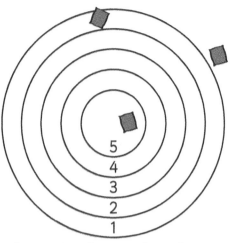

How many other solutions do you think Isla found? Could you have more than one bean bag missing the target? Why do you think this?

Let's reflect!

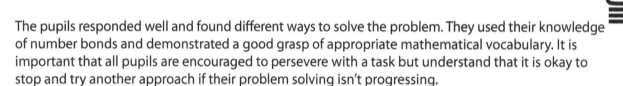

The pupils responded well and found different ways to solve the problem. They used their knowledge of number bonds and demonstrated a good grasp of appropriate mathematical vocabulary. It is important that all pupils are encouraged to persevere with a task but understand that it is okay to stop and try another approach if their problem solving isn't progressing.

The pupils generated their own questions while working on their solutions, which is hugely valuable and should be embraced. Set aside time to investigate questions arising at the plenary/sharing session. More time may need to be allocated to enable them to work on their questions as the pupils may want to explore further. It is important to capitalise on their interest and solutions.

Teacher Guide Reference

- Chapter 2 – Number and Number Processes – Number Recognition and Place Value
- Chapter 3 – Number and Number Processes – Four Operation

Bikes, Trikes and Scooters

Experiences and Outcomes

- *MNU 0-02a – I have explored numbers, understanding that they represent quantities and I can use them to count, create sequences and describe order.*
- *MNU 0-03a – I can use practical materials and can 'count on and back' to help me understand addition and subtraction, recording my ideas and solutions in different ways.*

Numeracy and Mathematical Skills

- **Interpret questions** – selects the relevant information
- **Select and communicate processes and solutions** – shares thinking
- **Justify choice of strategy used** – shows and talks through their thinking
- **Uses mathematical vocabulary** – uses developmentally appropriate mathematical vocabulary
- **Reason algebraically** – finds the unknown quantity

Resources

Access to bikes / scooters / trikes if the pupils are going to 'act out' the problem.

Before they start...

Pupils need to be able to count to 16 for the Extension activity.

The pupils need to know that, in this activity, bicycles have two wheels, tricycles have three wheels and scooters have four wheels.

Let's go!

Some bikes, trikes and scooters have been left outside instead of being put into the store.

Finlay counted and he spotted eight wheels altogether. Show the pupils pictures of the types of outdoor toys that are being referred to in the problem.

This picture shows one bike, one trike and one scooter; if we count the wheels, do you think we will get the same answer as Finlay? How can we work out what toys Finlay saw? Encourage pupils to think of a way of tackling the problem. **What should we do first? What could we use to help us? Can you remember how many wheels a scooter/bike/trike has?**

Enabling prompt

Talk through the problem and ensure that the pupils understand what is required. If the pupil is struggling with eight, reduce the number of wheels to four and see if they can identify the two possible solutions, i.e. two bikes or one scooter. For some it may be necessary to provide bikes/trikes/scooters to act out the situation. Can the children work out which toys were left outside by the number of wheels counted?

Extension activity

Use the same question as in the Let's Go activity but challenge the children to:

  a) find which group of toys would have 7 / 11 / 13 wheels or
  b) find how many different ways Finlay could have seen 7 / 11 / 13 wheels

Allow the pupils time and space to explore the activity using materials that they themselves think are appropriate for the task. The pupils may be able to draw wheels or use substitute objects such as cubes to help them solve the problem. Challenge them to find more than one solution.

Let's check!

Images are available for use on IWB or photocopiable resources.

Finlay counted 8 wheels.

There are a number of possible solutions, e.g.

- two scooters
- one scooter and two bikes
- two trikes and a bike

Pupils should work in pairs or work individually then share their thinking using 'think, pair and share'.

Extension activity

Finlay counted 12 wheels and then 16 wheels. There are a number of possible solutions:

12 wheels	16 wheels
3 scooters	4 scooters
1 scooter, 2 trikes, 1 bike	3 scooters, 2 bikes
6 bikes	8 bikes
4 trikes	3 bikes, 2 trikes, I scooter

Let's discuss!

Nuria's example of an 8-wheel solution

Nuria wasn't sure how to start. She understood that the toys had different numbers of wheels but didn't know what to do next. She looked at a scooter. She saw two wheels at the front and two wheels at the back. She knew that 2 + 2 = 4. She also knew that double 4 = 8. Nuria decided to check her solution and found that two scooters had eight wheels. **Did you choose two scooters for your solution? Did you get the same answer as Nuria? Did anyone find a different solution for eight? What did you use?**

Isla's example of a 12-wheel solution

Isla set out some counters. She set them in twos to represent the wheels on the bikes. She found that she had 6 lots of two to make 12. She did lots of counting and checking. She was counting in ones. **Was Isla correct? Was this what you found? Isla counted in ones; is there an easier way to count the counters Isla set out?** Isla then took 12 cubes. She put four cubes together to represent a scooter, three cubes to represent a trike and two cubes to represent a bike. She counted these and found that together that made nine. She had three cubes left. Isla realised that not only could the 12 wheels have been six bikes but it also could have been a scooter, a bike and two trikes. **Do you agree with Isla? Isla said four scooters made 12; how many toys did you use to make 12 wheels? What toys did you choose?**

Finlay's example of a 16-wheel solution

Finlay decided he would draw wheels to help him solve the problem. **Did anyone draw a picture to help them, like Finlay did?** He started by drawing 16 circles to represent his wheels. He found he could split the 16 wheels into 4 lots of 4. He drew four scooters to show his answer. **What did you do?** Finlay wondered how many bikes there might be with 16 wheels altogether. He drew two wheels and two wheels until he got to 16. He discovered it was eight bikes. He wasn't sure how many trikes there might be. He drew three wheels, then another three wheels and so on until he got to 15. He had five trikes but no toys had one wheel so he knew five trikes wouldn't work. Finlay wondered if four trikes would work and what else he could have to make 16. **Can you help Finlay? What should he do?** Finlay knew that two trikes had six wheels together; he knew 6 and 6 made 12. He took his 16 counters and set aside 12 counters for his four trikes. He had four left. **I wonder what Finlay might decide could make four? Can you think of any toys?** Finlay thought of his own bike, which has stabilisers on it. Finlay thought that would be a good solution.

Do you think we have found all the solutions? Did all the ideas we tried work?

Let's reflect!

From the solutions modelled, it is apparent that the pupils have approached the problem in different ways. There is evidence of effective use of manipulatives, repeated addition and using drawings to showcase their thinking. It is useful to further check, as a class, the possibility of any other solutions that may have been missed. Did anyone consider that there might be another option of a two- or three-wheel combination, e.g. a two- or three-wheeled scooter? Are the pupils confident in discussing their solutions and using their own words to describe what they did? Reflect on what mathematical knowledge the pupils may have used or are sharing; ensure that appropriate follow-up teaching is used, if required, to help progress their thinking, e.g. did the children count all or were they using counting on when they were exploring the total of wheels? Did they count in ones or did they skip count? Creating a table when collating data may help pupils to visualise all the options available.

Consider using a similar problem but in a different context. Look, listen and note whether the children can apply the learning from one context to another.

Teacher Guide Reference

- Chapter 2 – Number and Number Processes – Number Recognition and Place Value
- Chapter 3 – Number and Number Processes – Four Operations
- Chapter 11 – Data and Analysis

Broken Plates

Experiences and Outcomes

- **MTH 0-13a** – I have spotted and explored patterns in my own and the wider environment and can copy and continue these and create my own patterns.
- **MTH 0-16a** – I enjoy investigating objects and shapes and can sort, describe and be creative with them.
- *MNU 0-20b – I can match objects, and sort using my own and others' criteria, sharing my ideas with others.*

Numeracy and Mathematical Skills

- **Interpret questions** – selects the relevant information
- **Select and communicate processes and solutions** – shares thinking
- **Justify choice of strategy used** – shows and talks through their thinking
- **Uses mathematical vocabulary** – uses developmentally appropriate mathematical vocabulary
- **Reason algebraically** – finds the unknown quantity

Resources

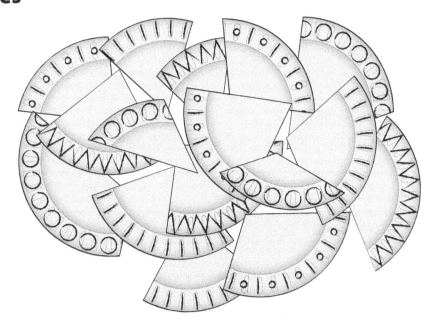

Paper plates marked with easily identifiable patterns/repeating patterns, each cut into four unequal parts (neither quarters nor regular pieces).

Before they start...

Pupils should be able to:

- identify and continue repeating patterns
- recognise common 2D shapes
- match and sort objects

Let's go!

Patterned paper plates (have prepared a number of plates similar to the ones in the picture).

Our plates have been cut up by mistake and we have a jumble of pieces. **I wonder what we can do. What do you think we need to do to get started? Can you estimate how many whole plates we would be able to make? What makes you think that?**

Enabling prompt

Provide two 'chopped up' paper plates (with two different patterns) per pair of pupils. Allow time for the pupils to talk about what they see and what they might do. Let them explore and investigate at their own pace. **What shapes can you see? How many different patterns can you see? Can you describe the patterns? Do you think they are repeating patterns?**

Extension activity

Increase the number of 'chopped up' paper plates per pair of pupils or add complexity to this task by either

- using similar plate patterns or
- having a plate with five pieces

Allow time for the pupils to talk about what they see and what they might do.

Let's check!

The solution will be to locate and place the appropriate pieces for each plate, while matching the pattern. The plates should be re-assembled to make a circle, the patterns should match and there should be no gaps. These questions can be used with the pupils to explore their knowledge, strategies and understanding of the problem:

- **Can you see a pattern? What do you notice? Can you see anything that is happening again and again? What shapes can you see? What shape is the plate? Can you recognise the shapes or markings on the plates?**

It is important to observe how the pupils start. The children may need to rotate the pieces to get the correct orientation. The pupils are best sitting side by side for this task so that they both have the same view.

- What do they notice / do first?
- Do they turn all the pieces over so the pattern is uppermost?
- Do they sort the pieces? How do they sort them?
- Do they check if there are four pieces with the same shapes / pattern?
- Can they visualise making a circle and getting the best 'fit'.
- Can they demonstrate sorting and matching?

Let's discuss!

Nuria and Finlay worked together. They had a pile of plate pieces in front of them. They knew that they had to make two plates from the pieces they were given. Finlay chose a piece with circles on it and Nuria chose a piece that had circles and lines on it. Nuria noticed that the circles on Finlay's piece were bigger than the circles on her piece. Together, they sorted the pieces into two piles: big circles and little circles with lines. They found it hard to fit the pieces together. It wasn't as easy as their jigsaw puzzles and they needed lots of tries. **What might have helped them?**

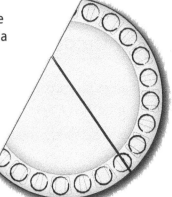

Amman and Isla had 12 pieces of plate that they sorted so that all the pieces were face up. They sorted them into patterns; Isla noticed one pattern that went circle, line, circle, line, and Amman spotted some zig zag lines that looked like triangles. They wondered how many whole plates they could make using the 12 pieces.
Amman took all the zig zag pieces and Isla took the line / circle pattern pieces. They tried hard to put the pieces together and with a bit of trial and error managed to make the plate whole again. They did the same for the other pieces and found that they could make four whole plates.

Let's reflect!

Nuria and Finlay solved the problem together. They managed to sort the pieces and make two plates using 'Guess and Check'. They didn't look for any patterns. If they had, they could have used this to find the next piece. Next steps for Finlay and Nuria could be to explore and complete a range of different jigsaws and puzzles to develop their understanding of flipping, turning or rotating pieces; a supportive adult should encourage and scaffold the use of appropriate maths vocabulary.

Amman and Isla also solved the problem. They were confident and capable when sorting. They recognised that some of the patterns were repeating, which helped them put the plate pieces together and they were able to describe, using appropriate mathematics vocabulary, the shapes used to make the patterns. They showed perseverance and were able to justify the choices they had made. They found it difficult to predict how many whole plates there would be.

Some pupils may, like Amman and Isla, have an interest in finding out how many whole plates could be made. This links to counting and sharing. Can they see the connections between four pieces per plate and the total number of plates? They could have shared out the pieces when sorting and found this out. This solution could be explored with all the pupils at the **'Let's discuss'** step using diagrams or the plates themselves.

It is always useful to ask pupils what they found easy or hard about the problem as this provides useful insight and can provide scope for developing next steps.

Some pupils may require a greater exposure to pattern and the application of pattern in different contexts. An emphasis on repeating patterns and predicting what comes next would be beneficial.

Teacher Guide Reference

- Chapter 8 – Patterns and Relationships
- Chapter 10 – Angles, Symmetry and Transformations

Car Park Sort

Experiences and Outcomes

- *MNU 0-02a* – I have explored numbers, understanding that they represent quantities, and I can use them to count, create sequences and describe order.
- *MNU 0-20b* – I can match objects, and sort using my own and others' criteria, sharing my ideas with others.

Numeracy and Mathematical Skills

- **Select and communicate processes and solutions** – shares thinking
- **Justify choice of strategy used** – shows and talks through their thinking
- **Uses mathematical vocabulary and notation** – uses developmentally appropriate mathematical vocabulary

Resources

- Selection of toy vehicles (including cars / lorries / buses, etc.) constructed of various materials (plastic / metal / wooden) with a number of different features or attributes such as number of wheels / number of windows / numbers marked on them / different colours, etc.
- A few large paper rectangles to represent the car parks (don't just limit it to two rectangles)

Before they start...

The pupils should be able to:

- recognise colours
- describe the size and shape of objects
- count up to 10 items

Let's go!

Have a big bag of toy vehicles (as suggested in Resources) for the pupils to explore. Say to the pupils, **I bought these at a car boot sale, can you sort them for me? What do you think we should do to get started?**

Enabling prompt

Have a smaller collection of vehicles with more obvious attributes (e.g. colour) and encourage the pupils to sort the vehicles using one criteria.

Extension activity

Look at how other pupils have sorted the vehicles and try to guess how they have sorted them (state the rule) or ask the pupils to sort the cars using two attributes.

Let's check!

There are many solutions to this activity that will depend on pupils' prior experiences. Sorting by colour / size / number of wheels may be the most common. Once the children have sorted the vehicles, ask them to choose how many car parks they require and justify their thinking?

Let's discuss!

Amman and Nuria decided to sort the vehicles by colour. They found that they had four colours of vehicles and needed four carparks: one for red vehicles, one for blue vehicles, one for yellow vehicles and one for white vehicles. They noticed when they were doing this that some vehicles had numbers on them (buses / racing cars) and decided that next time they would sort using a car park for vehicles with numbers and one for those that had no numbers marked on them. **Did you sort the vehicles the same way as Amman and Nuria? Which vehicles did you notice had numbers on them? Which car park had the greatest number of vehicles? Which had the smallest number of vehicles?**

Finlay and Isla decided to sort and re-sort using a different criterion each time. Finlay and Isla like counting so they chose to sort by looking at the number of wheels. As they did this, they discovered that most of the vehicles had four wheels. They labelled their carparks by drawing four circles for four wheels and lots of circles for more than four wheels. When they re-sorted, they sorted by the number of passengers that could travel in the vehicles. They had three car parks:

*   two-seater sports cars and lorries
*   family cars for four or five people
*   buses

Tell me how you sorted your vehicles. Did you manage to sort the vehicles in more than one way?

Let's reflect!

There is evidence that Nuria, Finlay, Amman and Isla were all able to make choices about how to sort the collection and could talk about the similarities and differences.

Ask the pupils – **Tell me what you did to get started. How did you sort the vehicles? Did you sort them in the same way or in a different way to Finlay and Amman? How did you know how many car parks you needed to have? Did you manage to sort the vehicles in more than one way?**

Next steps:

Can the pupils come up with suggestions of other ways to sort the collection of vehicles?

- Support the pupils to display their findings as a bar chart or pictogram instead of using paper rectangles; this may make the differences in the grouped items more obvious and develop data handling skills.

Teacher Guide Reference

- Chapter 2 – Number and Number Processes – Number Recognition and Place Value
- Chapter 11 – Data and Analysis

Dandelion Clocks

Experiences and Outcomes

- *MNU 0-01a – I am developing a sense of size and amount by observing, exploring, using and communicating with others about things in the world around me.*
- *MNU 0-02a – I have explored numbers, understanding that they represent quantities, and I can use them to count, create sequences and describe order.*

Numeracy and Mathematical Skills

- **Interpret questions** – selects the relevant information; interprets data
- **Select and communicate processes and solutions** – shares thinking
- **Determine the reasonableness of a solution** – routinely uses estimation

Resources

- A collection of dandelions in the 'clock' phase of their cycle
- Tablet to record
- IWB picture of a dandelion
- Copy of Dandelion Clock poem
- Resources to help count the score, e.g. chalk, pebbles, bead strings, sticks

Before they start...

Pupils should have experience of:

- counting and keeping track of their count (the number range depends on how hard the pupil can blow and the size of the dandelion head (extension activity))
- recording their findings (using their own system)

Let's go!

Read the poem Dandelion Clock (see downloadable resources). Explain to the pupils that this is an old tradition or folktale that has been passed down from generation to generation. **Have you ever tried using a dandelion clock to tell the time? What did you do?** Ensure the pupils realise that this is not an accurate means of telling the time but that it is a fun activity that can help spread the seeds for the dandelion plant. The activity may also provide a context for discussion about the changing seasons. Consider filming some pupils investigating this problem on a tablet so that it can be used throughout the year.

Timelapse videos are available online that provide a visual of the dandelion changing into a dandelion clock.

Allow the pupils time to explore outdoors and search for dandelion clocks. Discuss and compare the plants collected. **Are all the dandelion clocks the same size?** Encourage the use of comparative language such as more / less, big / little, bigger / smaller. **How many puffs do you think it will take to blow all of the seeds off this dandelion head? How will you keep track of your count?**

Enabling prompt

Ensure that the pupils have had lots of opportunties:

- to collect and count small collections, using a wide range of practical materials, e.g. pine cones / buttons / washers / cubes / pebbles
- to use key vocabulary, e.g. less than / more than or the same as
- to practise forward number sequences up to 10 / 20
- to compare collections
- to estimate by looking, e.g. at pictures or illustrations where they can't actively move the collections to compare

Extension activity

Pupils should carry this activity out in pairs; encourage them to predict the number of puffs needed for a wide range of dandelion clocks, to order their predictions from most to least and to choose their own method of recording the count. **What did you notice? What surprised you? Did it take longer to blow the seeds off the bigger dandelion heads? Always? Why do you think that happened? Apart from size, what else is important?**

Let's check!

This is an activity that can be repeated many times and that the pupils can explore themselves when playing freely. The variables are discussed above (size of dandelion clock / amount of puff). As this is a seasonal activity, you may wish to film the pupils and refer to the recordings out of season.

Let's discuss!

Nuria and Isla compared the sizes of two dandelion heads by holding them side-by-side. They noticed that one dandelion had a much larger head than the other. **Do you think they will need the same number of puffs to blow all of the seeds off each dandelion head? Can you explain why you think that?** Nuria puffed and Isla counted on her fingers. Isla was stuck when Nuria got to 10 puffs as she had no more fingers to record on. **Did anyone else count on their fingers like Isla? What could Isla do next?** Isla then had a turn at puffing and Nuria recorded. Nuria put a pine cone into a tub every time Isla puffed and counted the pine cones at the end. Isla liked Nuria's idea and said she'd use it next time. **How did you record your count?**

Amman noticed that the dandelion clock he had chosen had lost a few seeds . . . he wondered if the gaps would be worth one or two puffs. **What do you think?** He thought he might add this on at the end of his count. **Do you think that is fair? Why/why not?**

Finlay chose one of the biggest dandelion clocks. He estimated that it would take 12 puffs; he asked Isla to help him count the number of puffs he used. **He asked her to move a bead along a bead string everytime he puffed some seeds off. What did you use to record your count? How close was your estimate?**

Let's reflect!

Recording can be as simple as chalking a mark each time a 'puff' is made, putting a pebble into a pot or keeping track on their fingers and writing down the total number of puffs. If using a recording of the activity, the pupils can be encouraged to keep count on their fingers. The 'ripeness' of the dandelion head alters the ease at which the seeds are dispersed. Try comparing the clocks for size before the activity starts; roundest / biggest at one end of a line to thinnest / smallest at the other end of the line. Did the largest dandelion head take the biggest number of puffs? Ask the pupils to come up with ideas of how to record their results, collate and display the data. Listen for the pupils using mathematical vocabulary.

From the solutions provided, it is clear that some deep thinking was occuring, e.g. Amman recognised that a dandelion clock wth seeds missing may not give a true count. His thoughts and findings provide valuable discussion points and should be used to progress the pupils' thinking.

Teacher Guide Reference

- Chapter 2 – Number and Number Processes – Number Recognition and Place Value

Exploring Patterns

Experiences and Outcomes

- **MTH 0-13a** – I have spotted and explored patterns in my own and the wider environment and can copy and continue these and create my own patterns.

Numeracy and Mathematical Skills

- **Selects and communicates processes and solutions** – shares thinking; verbalises or demonstrates thought processes
- **Justify choice of strategy used** – shows and talks through their thinking
- **Link mathematical concepts** – transfers learning in one area to another
- **Use mathematical vocabulary and notation** – uses developmentally appropriate mathematical vocabulary

Resources

- Selection of pattern starter cards
- Selection of resources for pattern making, e.g. buttons, pom-poms, loose parts (e.g. washers, screws, pine cones, shells), counters, coins, pasta, shapes, etc.
- Pattern copy cards (see Enabling prompt)
- Threading beads and lace / string (see Extension activity)

Before they start...

Pupils should understand what a pattern is and have experience of:

- copying repeating patterns
- continuing repeating patterns
- creating repeating patterns
- describing patterns / repeating patterns

Let's go!

Ask the pupils to choose a pattern starter card. Tell them you would like them to use the pattern that is on the card to make a repeating pattern. You do not need to choose the shapes that are on the card – you may pick other objects from the pattern-making box. For example:

What do you notice about this pattern? What will the next shape be? And the next shape? What could you use instead of a triangle / square? How many of each will you need? Why?

Encourage the pupils to choose from a range of manipulatives and materials to create their repeating pattern; they may take a photograph of their pattern using a tablet.

Talk to the pupils about their patterns. **What have you used to represent the triangle, square, etc.? Why did you choose that? How many times does your pattern repeat? If you were to continue your pattern further, what shape would come next?**

Enabling prompt

Provide the pupils with pattern copy cards for them to explore, talk about and copy.

Copy this pattern. What colour comes next?

e.g.

AND/OR

Play 'repeating or not'. In this game, the pupils are shown pictures of patterns (IWB) and are asked to respond using actions, e.g. arms up for repeating pattern or crossed arms for not a repeating pattern.

Yes   No

Extension activity

Challenge the pupils to make their own repeating pattern necklace using threading beads. The pattern on the necklace must continue when the necklace is 'fastened'. Pupils can be asked to predict how many of each bead they will need before they begin.

Look at all of the patterns created. Ask the pupils to identify whether the pattern is repeating and where it repeats.

Pupils could also be challenged to use a sequence of numbers to represent a repeating pattern, for example:

can be represented by the sequence 2, 1, 2, 1, 2, 1.

Let's check!

In the enabling prompt, the pupils are being asked to copy a pattern; in the main problem, the pupils are asked to copy and continue the pattern. Scaffolding may be required to enable some pupils to describe their repeating pattern using mathematical vocabulary. It is easier for a pupil to talk about the pattern they have made when they have it in front of them. Look out for inconsistency in copying or extending the repeating pattern unit.

Let's discuss!

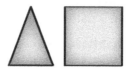

Finlay chose this pattern starter card.

He knew he would have to have two different objects; he chose big pom-poms and small pom-poms.

Here is his pom-pom AB pattern – big, small.

Did you have a starter card like Finlay's? What did you use to make your AB pattern? Has Finlay finished making his repeating pattern? How can you tell?

Amman chose a pattern starter card but didn't know what to do as he couldn't find any triangles or squares to copy the pattern.

Finlay said it didn't need to be the exact same objects as on the card as long as he had two different objects and had the correct number of each. Amman chose pasta shapes. **Is Amman's pattern correct? How many pasta bows did Amman use altogether?**

Amman ran out of space to finish his repeating pattern. **What pasta shape does he need to finish it?**

Nuria decided to make her own ABC pattern. This is the pattern she made using nine triangles.

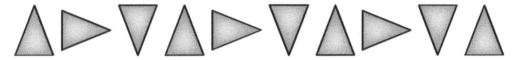

Do you think Nuria's pattern is a repeating pattern? What did she do to make her pattern? Did you make your own pattern? How did you know what came next?

Isla used number cards to make this repeating pattern.

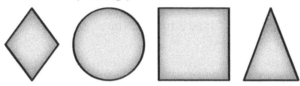

She needed four different cards.

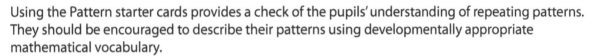

Is Isla's pattern correct? What makes you say that? Do you think the numbers should be in order in her pattern? Why / why not? Did you use numbers to make a repeating pattern? Tell me about your pattern.

Let's reflect!

Using the Pattern starter cards provides a check of the pupils' understanding of repeating patterns. They should be encouraged to describe their patterns using developmentally appropriate mathematical vocabulary.

Some pupils may have difficulty in recognising that the example shown on the starter card is representative of the pattern to follow; it can be made using many different manipulatives and not just the shapes depicted. Using practical materials that the pupils are already familiar with can reduce the cognitive load placed upon them and enable them to explore and focus on pattern creating.

Ensure that pupils have opportunities to explore a range of repeating patterns both indoors and out, including patterns in their environment.

Teacher Guide Reference

- Chapter 2 – Number and Number Processes – Number Recognition and Place Value
- Chapter 8 – Patterns and Relationships

Finding Half

Experiences and Outcomes

- *MNU 0-07a – I can share out a group of items by making smaller groups and can split a whole object into smaller parts.*

Numeracy and Mathematical Skills

- **Select and communicate processes and solutions** – shares thinking; verbalises or demonstrates thought processes
- **Justify choice of strategy used** – shows and talks through their thinking
- **Use mathematical vocabulary and notation** – uses developmentally appropriate mathematical vocabulary

Resources

Four or five boxes of objects / materials to investigate. Use items that involve measuring, cutting and comparing.

- Resource sheet – Finding Half: Halves mat

Suggested materials / objects to explore:

- Dough, plastic knives and pan balances

- Transparent plastic tumblers, jugs and water or coloured liquid

- String / wool, strips of paper and scissors
- Paper shapes (different shapes and sizes)

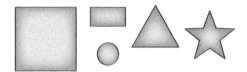

Before they start...

Pupils should, through play and real-life experiences, be familiar with the words half, fair share and equal parts.

Let's go!

Finding half.

Give the pupil a box containing one item, e.g. a lump of dough, a strip of paper, a paper shape, a jug of sand / liquid.

Nuria wants to share what's in the box with Isla so that they have half each. Can you help her? Allow the pupils time to explore. Observe what they do and challenge them to justify their thinking and decision making. **What did you do to half what was in your box? Do Nuria and Isla have half each? How do you know? How could you check?**

Enabling prompt

Provide, through play, opportunities to fairly share out objects e.g. setting a table in the picnic area. **Do we have enough plates / teaspoons / cups for everyone to have one of each?** Also include activities that involve exploring mass, such as ensuring that dry sand is shared equally into two buckets; provide balance scales to check whether the amounts of sand are equal.

Question pupils to elicit the extent to which they are aware that in order for each piece to be 'half' there needs to be two of them and that they need to be the same size / amount / mass / capacity. **When we split something into two equal parts, what do we call each part? Is it still half if the parts are not equal?**

If necessary, ask, e.g. **Can you think what we might do to split the dough into two pieces of dough that are the same size? What might help us to get started? How might we share the water in the jug so that there is the same amount in each glass? What could we do to make two equal-sized pieces of string from our original piece? What would we need to do to split the strip of paper into two equal-sized pieces? What should we do to split the shape into two equal parts?**

Extension activity

Provide the pupils with a box containing more than one item.

Finlay wants to share what is in his box with Amman so that they each have a half. Allow the pupils time to explore. Observe what they do and challenge them to justify what they did to share the objects fairly.

Provide boxes / containers and materials for sharing, e.g. beads and bead strings, pom-poms, pennies, counters, etc. Have available paper plates / bowls or cups to share the items into if required.

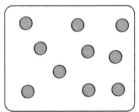 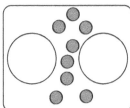 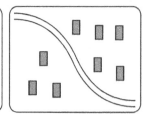 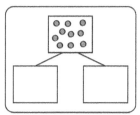

This Extension activity of finding half is not solely about the pupils sharing 'one to me and one to you' but about them developing an understanding that by making two equal-sized groups, they have halved the collection. Discussion should focus on equal groups / shares. Encourage the pupils to check that groups / shares are the same size by counting and / or matching.

Can you tell me what you did? How do you know that you have halved your collection? Were there any objects left over? Can they be halved?

Let's check!

Depending on the context, pupils may use one of the following to solve the problem: counting, equal sharing, comparing, measuring, folding or cutting.

Solutions Main Task

Dough

Do the pupils:

- split / cut the dough into two unequal pieces?
- split / cut the dough into two pieces that look similar in size?
- use a pan balance or other means of weighing to compare the two pieces?

Paper shapes

Do the pupils:

- cut the shape without first checking that the two pieces will be the same size?
- fold the shape over but not down the mid-line and do not check that the parts are equal?
- fold the shape over on top of itself, so that both parts are the same size?

String, wool or strip of paper

Do the pupils:

- compare the two pieces to check that they are the same length before cutting?
- measure the two pieces (using non-standard units, e.g. paperclips) to check that they are halves after cutting?

Solutions Extension Task

Counters / pennies

Do the pupils:

- split the collection without counting (push counters into two separate bundles)?
- share the collection without counting ('one to me, one to you')?
- count the total collection and then split it into two equal parts; the children may match, using one-to-one correspondence to check that the collections have the same amount in each?

Cubes

Do the pupils:

- move the cubes into two bundles without counting?
- share the cubes out into two bundles, then count to see if they have the same number in each bundle?
- build the cubes into towers and compare?
- build a tower of cubes and split the tower into two equal towers?

Beads and strings

Can the pupils make two necklaces or bead strings of equal length?

Do the pupils:

- thread the beads onto the two strings and then measure the strings?
- thread the beads onto the two necklaces and then count?
- share out the beads into two equal groups and then thread?

Pom-poms and plates

Can the pupils share the pom-poms equally between the two plates?

Do the pupils:

- put some pom-poms onto each plate by comparing the collections, rather than counting?
- share the pom-poms out without counting the collection first – 'one to me and one to you'?
- count the total collection and then split into two equal amounts on the plates?

Halves mat

Provide a picture of a halves mat already populated. Can the pupils predict what number of counters would be half of the collection?

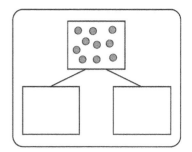

Let's discuss!

Finding half:

Isla and Nuria chose a box that had some dough in it. They pulled the dough apart so that each of them had one piece. They thought the pieces of dough looked the same size. **I wonder if they have found halves. Can you think of a way to check that the two bits of dough are the same size and are halves?**

Halving a collection:

Amman and Finlay opened their box and found a picture of a halves mat. **Can you think of a way to get started?** Finlay thought it would be a good idea to place cubes on top of the circles in the top box and then share them out into the two boxes below. **Do you think this is a good way to start the problem? Why do you think that? What would you have done?**

Let's reflect!

Linking everyday events and experiences of sharing will make situations relevant to the pupils, e.g. giving out classroom resources, sharing out fruit at snack time or organising the pupils into groups for activities. This introduces the notion of fairness and equality; modelling the use of sharing and grouping supports the pupils to develop a secure understanding.

Pupils need lots of opportunities to talk about, explore, compare and physically group and share whole objects and amounts. Do the pupils recognise that if they split something into two parts / pieces they may not necessarily be creating halves? The pupils should recognise that the two parts / pieces need to be of equal size to be classed as halves. Vocabulary used can tell us much about the pupil's understanding. It is important to explore, in context, terms such as whole, half, halves, fair, share, group, left over.

When exploring halving shapes of different sizes, it is important that the pupils develop the understanding that the size of the half depends on the size of the whole and it may be possible to halve the same shape / object in different ways.

Teacher Guide Reference

- Chapter 4 – Fractions

Loose Parts 1 – How much am I?

Experiences and Outcomes

- *MNU 0-09a – I am developing my awareness of how money is used and can recognise and use a range of coins.*

Numeracy and Mathematical Skills

- **Select and communicate processes and solutions** – shares thinking
- **Link mathematical concepts** – transfers learning from one area to another
- **Use mathematical vocabulary and notation** – uses developmentally appropriate mathematical vocabulary

Resources

- Resource sheet – Loose Parts 1 – How much am I?
- 'Catalogue' of outdoor loose part portraits (including photographs of the 'portraits' created in 'Making Faces – Loose Parts 2' if pupils have tried this problem)
- Price List (can be adapted to suit the pupils' ability and the items available) e.g.

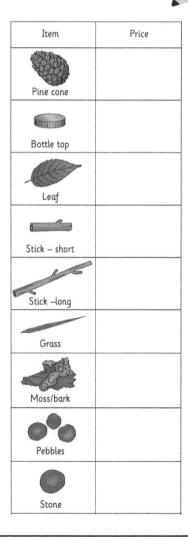

Item	Price
Pine cone	
Bottle top	
Leaf	
Stick – short	
Stick –long	
Grass	
Moss/bark	
Pebbles	
Stone	

Item	Price
leaf	1p
pebble	2 for 1p
stone	1p
pine cone	2p
small stick	2p
large stick	5p
grass	1p
bottle top	1p
dandelion	2p
Bark/moss	1p

- Coins (purses if necessary to keep coins together)
- Grid for recording
- Blank price tags and sticky tape or glue

Item	Cost (How much?)	How many used?	How much altogether?
Pine cone			
Bottle top			
Leaf			
Stick – short			
Stick –long			
Grass			
Moss/bark			
Pebbles			

- Pencils / dry-wipe pens / wipe-clean whiteboards
- Empty 10 frames

Before they start...

The pupils should:

- have experience of using coins to pay for goods in a role-play situation
- be able to recognise and name coins, e.g. 1p, 2p, 5p, 10p

Let's go!

Show the pupils the price list and talk to them about what could be bought in the shop. Ask the pupils if they think any other items should be included on the price list.

Show the pupils these portraits (or if they have completed the Making Faces task, use one of the faces created previously).

Look at these pictures. How could you work out the cost of buying the things we need to make this portrait? Do you think your total will be more or less than 10p? Why/why not?

Use coins to find the total cost of each picture in our 'catalogue'. See downloadable resources.

Enabling prompt

For pupils who are having difficulty, provide:

- a portrait that doesn't use lots of items
- coin strips for relevant items rather than using the pricing grid (for the children to place their 1p coins onto)

Item	1p coins
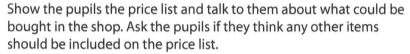 Bottle top	

- a sufficient number of 1p coins rather than having to use 2p or 5p coins, e.g. a large stick costs 5p – encourage the children to count out the correct number of 1p coins and find the total by counting.

Did the portrait you are looking at need one of everything? Which thing did you need most of / least of? Which thing cost most / least? Which costs more, a cone or a leaf, etc?

Extension activity

Once the pupils have calculated how much the portraits are worth, they should write a price tag for each portrait and arrange them in order from cheapest to most expensive. Alternatively, have ready-made price tags with the totals written on for the pupils to match to the portraits.

To provide even greater challenge, have different prices for different coloured bottle tops and colours / sizes of pebbles. Encourage the pupils to help set the prices.

Let's check!

There are multiple possible solutions to this problem, depending on the portraits chosen and the prices used. You will need to check the total price against the items on the portrait. For the extension task, the price tag should match the total cost of the portrait.

Let's discuss!

Isla chose this face. She thought it looked like her dad as it had a beard.

Isla completed the table.

Item	Cost (How much?)	How many used?	How much altogether?
Pine cone	2p	2	4p
Bottle top	1p	2	2p
Leaf			
Stick – short	2p	4	8p
Stick –long			
Grass			
Moss/bark	1p	1	1p
Pebbles			
Stone	1p	1	1p
Dandelion	1p	1	1p

Isla said it was a lot of numbers to add up. She decided that she would use 1p coins to help her add up the cost of the portrait. She counted out the correct number of 1p coins and placed them onto two empty 10 frames, beside the grid. She thought the total price was 17p. **Is Isla correct? How can we check? Why do you think it was a good idea to use 1p pieces? Show me your portrait. How did to work out how much it cost?**

Let's reflect!

Pupils at this stage need lots of opportunities to handle coins in 'shopping' contexts. Not all will appreciate that, for example, one 2p coin has the same value as two 1p coins or that five 1p coins have the same value as one 5p coin. They may say that 2p + 1p + 1p + 2p + 1p coins = 5p because there are five coins. Encouraging them to count out the equivalent number of 1p coins to 'match' the numeral on the coin can help. Empty ten frames can be provided and pupils guided to place the 1p coins onto them to help with the final count. **How much would a full 10 frame of 1p coins be worth? Which coin would match this amount?** For pupils unable to work out the total cost, attach price tags to ready priced portraits for them to order.

Teacher Guide Reference

- Chapter 5 – Money

Musical Money Boxes

Experiences and Outcomes

- *MNU 0-03a* – *I use practical materials and can 'count on and back' to help me understand addition and subtraction, recording my ideas and solutions in different ways.*
- *MNU 0-09a* – *I am developing my awareness of how money is used and can recognise and use a range of coins.*

Numeracy and Mathematical Skills

- **Interpret questions** – selects the relevant information
- **Select and communicate processes and solutions** – explains choice of process; shares thinking
- **Justify choice of strategy used** – shows and talks through their thinking
- **Link mathematical concepts** – transfers learning in one area to another
- **Uses mathematical vocabulary and notation** – uses developmentally appropriate mathematical vocabulary
- **Mental agility** – knowledge of number facts

Resources

- Instructions for the Musical Money Box game (see Resource sheet)
- Access to concrete materials of their choice including access to coins (1p, 2p, 5p and 10p)
- Blank ten frames
- IWB images of coin cards (1p, 2p, 5p and 10p)

Before they start...

The pupils should be able to:

- recognise coins (1p, 2p, 5p and 10p)
- add within 10

Let's go!

Play Musical Money Boxes with your class / group of pupils (see instruction sheet). Look closely for pupils who have yet to enter a box. Ask **What coin do you have? Can you see any empty boxes that your coin card matches?** If there are none, say, **Look at the other children already in their boxes, holding their cards up. Could you join this box? Why / why not? Can you see a box you might be able to join?** If there is no possible solution, the pupil is out but becomes a 'money guru' and provides support to others.

Note:

- What was problematic for some pupils?
- Were there any coin combinations that they found tricky?

Gather pupils together and discuss what they discovered/learned while playing the game. Ask, **Which amounts did you match? Did you join up to make a total? How did you work out the total? Did you talk to a partner to find the total? What did you say to each other?**

The following games can be used with the pupils to extend their thinking.

Game I

Nuria, Finlay, Amman and Isla were playing Musical Money Boxes in the playground. They skipped around the chalked boxes, carrying their chosen coin cards. When the whistle blew, Amman and Isla rushed to the 6p Money Box. Finlay and Nuria looked at the cards that Amman and Isla were holding up and joined them in the box. They were certain that the total of their coins was correct for that box. **I wonder what coin cards they each had. Could they all have had coin cards of the same value? What makes you say that?**

Game 2

Display the image below on an IWB. Say, **Look carefully at the value written in each box and the cards that the four children are holding. Which boxes could the children go into? Who can only go into one box? Who can go into more than one box? Which boxes could Nuria enter? etc.**

Enabling prompt

Provide the pupils with their coin values in 1p pieces to help them compare amounts and add the totals together. The value of each Money Box could be shown as a 10 frame with 1p coins in it, e.g.

Extension activity

One of the chalk labels for the money box has been rubbed off. Can you think what the value might have been? Here is a clue to help you: **there is one Money Box that all four pupils can fit into.**

Provide challenge for the pupils by using Games 3 or 4 or increase the values marked in the boxes.

Game 3

Show the pupils Game 3 on an IWB.

The four children were playing Musical Money Boxes. Nuria, Finlay and Isla all ran into one of these money boxes. Amman wasn't able to join them but he was able to go into one of the other money boxes by himself. **I wonder which money box Nuria, Finlay and Isla ran to. Which coin cards do you think they each had? Which coin card do you think Amman had? Can you tell me which box he might have been able to go into? Why do you think that?**

All four children would like to be in the same box. Which coin cards will allow them to do this?

Game 4

Show the pupils Game 4 on an IWB.

Amman and Finlay ran into a money box together. Nuria and Isla ran into two of the other boxes. **What coin cards do you think they all had? Can you think of a combination of coin cards which could go into the other box? Can you tell me why you think that?**

Let's check!

Solutions

Game 1 – Money box total is 6p. The coins that the characters have are: 2p, 2p, 1p, 1p.

Game 2 – The total of their coins would be 18p. There is no box that the four children can enter together. Finlay can go into the 10p box alone. Isla can go into the 5p box alone. Isla can go into the 8p box with Nuria and Amman. Nuria and Amman could go into the 3p box together.

Game 3 – There are a few possible solutions to this problem depending on the coins each child had. Finlay, Nuria and Isla, together, could enter the:

- 3p box if they each had a 1p coin (Amman could have a 2p coin, a 5p coin or a 10p coin)
- 5p box if they had a combination of 2p, 2p and 1p coins (Amman could have a 1p, 2p, 5p or a 10p coin)

All four children would be able to enter the 10p box if they had 5p, 2p, 2p and 1p cards.

Game 4 – Amman and Finlay went into the 7p box; one had a 2p coin and the other had a 5p coin. Nuria had a 2p coin and went into the 2p box. Isla had a 10p coin and went into the 10p box. The empty box had a value of 9p – two possible coin combination are 5p and 2p and 2p or 5p and 1p and 1p and 2p.

Let's discuss!

Game 1

Amman and Isla both had 2p coin cards; they knew that together their coin cards made 4p. While skipping round the boxes they were looking for a Money Box with a value of 4p or greater. They ran into the 6p money box. Finlay and Nuria joined them in the box. **What coin card do you think Finlay and Nuria both had? Why do you think that?**

Game 2

Finlay had a 10p coin card. He knew his coin meant he could only go into one box and he also knew he couldn't share the box with anyone else. **Can you think which box that might be?** Isla saw that there was a box that she could go into alone. **Which box could that be?** Amman and Nuria wanted to go into boxes by themselves but they weren't sure if they could. **What do you think?** The children wondered what amount would need to be written in a box if they were all to be able to go in it together. After they had played the game, they got out some empty ten frames. They swapped their coin cards for 1p pieces and placed them in the ten frames to help them count how many. They had one full ten frame and a ten frame with two empty boxes. **How much do they have altogether? How did you work it out? Can they all go into a box together? Which one?**

Game 3

Isla, Nuria, Finlay and Amman found that they each had been given a coin card with the same value. The girls knew that 2p and 2p made 4p. They were looking for a box that had a 4p or more marked on it. They ran into the 10p box and Finlay ran in after them. **Can Finlay stay in that box? Why not?** Amman ran to another box and was able to stay by himself. **Which box might that be? What coin cards did the four children have? Could we try this problem again using different coin cards?**

Game 4

Amman and Finlay looked at their coin cards and they knew that they could go into the 7p box together. **What coin cards do you think they had?** Nuria and Isla ran into two of the other boxes. **Can you work out which boxes Nuria and Isla ran into?** Allow some thinking time then prompt by saying **Isla had a 10p coin. What coin do you think Nuria had? Which money box had no children in it? What combination of coin cards could be used for that box? Why do you think that?**

Let's reflect!

It is important that pupils have concrete materials available, e.g. 1p coins and empty 10 frames to help support their counting. They should be encouraged to demonstrate and compare strategies for working out the totals. Initially, the pupils may experience the game physically and then progress to solving the problems using pictorial representations.

An equivalence chart may be useful in supporting some learners to see coin combinations for particular values, e.g. how many different ways can we make 5p? It is important that the pupils understand that different collections of coins can have the same value.

Teacher Guide Reference

- Chapter 2 –Number and Number Processes – Number Recognition and Place Value
- Chapter 3 –Number and Number Processes – Four Operations
- Chapter 5 –Money

Help the Park Keeper!

Experiences and Outcomes

- *MNU 0-02a* – *I have explored numbers, understanding that they represent quantities, and I can use them to count, create sequences and describe order.*
- *MNU 0-03a* – *I use practical materials and can 'count on and back' to help me understand addition and subtraction, recording my ideas and solutions in different ways.*
- *MNU 0-20a* – *I can collect objects and ask questions to gather information, organising and displaying my findings in different ways.*

Numeracy and Mathematical Skills

- **Interpret questions** – selects the relevant information; interprets data
- **Select and communicate processes and solutions** – shares thinking
- **Justify choice of strategy used** – shows and talks through their thinking
- **Use mathematical vocabulary and notation** – uses developmentally appropriate mathematical vocabulary
- **Mental agility** – knowledge of number facts

Resources

- Resource sheet – Help the Park Keeper
- Toy zoo animals
- Paper and drawing tools (pen / pencil / crayon)
- Fiction or non-fiction books about safari / zoo animals

Before they start...

The pupils should be able to count to up to 12 for the initial story and beyond for those attempting the extension.

Let's go!

Read this story:

It was a stormy, wet winter at the safari park and the park keeper was having trouble getting across to the island, to feed the animals that lived there; he decided to move them to different enclosures in the main park. A boat was organised for the animals to travel in. As they walked onto the boat, the park keeper counted 12 legs. What animals could he have seen?

It is important to ensure that the pupils have grasped the important information from the story.

What is the story telling us? What do we need to find out? What would help us solve the problem? Have you done anything like this before? I wonder what animals the park keeper saw. What should we do first?

Brainstorm and note any suggestions the pupils make. A list of safari park animals could be made to help the pupils visualise the problem or have a range of picture books for the children to use as reference.

Enabling prompt

The following are suggestions to help pupils get going with solving the problem.

The pupils can undertake some research before tackling the problem; they may need information on the types of creatures that live in a safari park.

Re-read the story, if necessary, to ensure that the pupils recognise the important information they will need to find a solution.

Allow the pupils time and space to explore the activity using materials that they themselves think are appropriate for the task. They may choose to use plastic zoo animals or look at pictures of animals for ideas.

You can provide a list of animals and the number of legs each has. This is a prompt to help the pupils get started. Do not make an exhaustive list as this may hinder creativity.

Animal		Number of legs
Bear		4
Tiger		4

Animal		Number of legs
Giraffe		4
Eagle		2
Flamingo		2

Extension activity

Challenge the pupils to find more than one solution. If they find multiple solutions for 12, then increase the number of legs. Ask the children, **Do you think there might be an odd number of legs?** Note that the information from the enabling prompt is still relevant.

Allow the pupils time and space to explore the activity using materials that they themselves think are appropriate for the task. The pupils can choose to draw animals or use substitute objects such as counters to help them solve the problem.

Let's check!

There are many possible solutions. It is important for the pupils to see that there are many different ways of making 12.

12 legs: three lions; three bears; two lions and a bear; two eagles, a lion and a giraffe; etc.

16 legs: two bears and two giraffes; two flamingos, a tiger, a giraffe and a bear; etc.

How many ways did you find to make 12 legs? What animals did you use? Do you think these are the only ways to solve this problem? Can you tell me more?

Let's discuss!

Here are some solutions found by four other children. I wonder if anyone here had the same ideas as them?

Amman collected some plastic zoo animals. He chose his favourite ones – a lion, a tiger, a giraffe and a bear. He said that these four creatures would have 12 legs altogether. **Is Amman correct? How do we know? How many legs do you think there are in total?**

Isla likes penguins and tigers. She knows that penguins have two legs and tigers have four legs. Isla has been working on doubling numbers and knows that $2 + 2 = 4$. So, two penguins together would have four legs altogether. Isla knows that tigers have four legs; she knows double four. So, two tigers would have 8 legs altogether. She knows $8 + 4 = 12$ so there could be two penguins and two tigers on the boat. **Did anyone find a solution like Isla's? Can you tell us what you did?**

Nuria likes spiders and she knows that they have eight legs. She wonders how many more legs it would take to make 12. She counts back from 12 to 8, decides it is a creature with 4 legs and chooses an alligator. **How would you have solved Nuria's problem? Can someone tell me how you would have solved it?**

Finlay had been to a safari park. He thought about what he had seen. He remembered seeing the big brown bears. He had seen them walking on two legs and on four legs. Finlay did some deep thinking and he decided to draw a picture to show his thinking. He drew two bears walking on two legs (four legs altogether). He then drew three bears walking on four legs (12 legs altogether).

He then counted all the legs – it made 16. **Did you find a solution similar to Finlay's? Do you think you would have used a count all strategy like he did? How else might Finlay have worked out the number of legs?**

Let's share our ideas. How many different ways have we found to solve the problem? Are any of our solutions the same as Finlay, Nuria, Isla and Amman's?

Let's reflect!

Amman needs support to select the relevant information. He started well by choosing zoo animals to support his investigation but forgot to count and check the number of legs for four animals. Reminding Amman of the question may have helped him complete the final step. He only selected some of the relevant information – zoo animals and the number 12. He guessed but didn't touch count to check his answer. He found it hard to justify his answer. By sharing solutions, Amman will have the opportunity to hear and see other solutions.

Isla is able to select the information from the story. She has used and connected previous knowledge and has applied it in a new context. She realised that using a counting on strategy and fingers would help her to solve the problem and check her solution.

Nuria has shown application of her numeracy knowledge. She was able to extract the relevant information from the story and solve the problem. She wondered about another solution and thought two spiders would make 16. She drew two spiders and counted the legs to check.

Finlay has used previous knowledge and creativity to provide a lovely solution to the question. He was able to justify his solution by showing his pictures; challenge was supplied by increasing the number of legs.

The pupils are showing confidence in working through problems. Amman might be struggling slightly and needs some support in listening and processing the information; he may have benefitted from a pictorial reminder. It is important, when providing support, not to give a pupil the answer but encourage them to persevere. By modelling and using brainstorming, list making, drawing pictures and encouraging guess and check it is possible to provide children with some 'tools' for their problem-solving toolkit; pupils should experience a range of different ways to process information and solve problems. This problem has a similar approach to Bikes, Trikes and Scooters but uses a different context. It would be interesting to see if any of the pupils make links between them.

Teacher Guide Reference

- Chapter 2 – Number and Number Processes – Number Recognition and Place Value
- Chapter 3 – Number and Number Processes – Four Operations

How Far Did You Get?

Experiences and Outcomes

- *MNU 0-10a* – *I am aware of how routines and events in my world link with times and seasons, and have explored ways to record and display these using clocks, calendars and other materials.*
- *MNU 0-11a* – *I have experimented with everyday items as units of measure to investigate and compare sizes and amounts in my environment, sharing my findings with others.*

Numeracy and Mathematical Skills

- **Select and communicate processes and solutions** – shares thinking; verbalises or demonstrates thought processes
- **Use mathematical vocabulary and notation** – uses developmentally appropriate mathematical vocabulary

Resources

- Resource sheet - How Far Did You Get?
- Selection of sand timers / electronic timers
- Chalk (thick playground chalk is best)
- Roll of backing paper and marker pens if the pupils are working indoors.
- For Extension activity – selection of non-standard units to measure the lines drawn

Before they start...

The pupils should have experience of:

- using sand timers to record the passage of time
- drawing lines with chalk
- measuring length using non-standard units (extension activity)

Let's go!

Mark a cross on the playground as the starting point for this activity. Ask the children to predict: **How far do you think you might be able to walk to, from here, before the timer runs out? Start with 30-second timer and progress to 1 minute; alter the timer used to suit the pupils involved.** Have available bean bags / stones / pine cones, etc. for the pupils to place at the distance they think they might reach. **Explain to the pupils they will be making a long chalk line, as they walk,** from the X. **Do you still think you will be able to walk that far? What makes you say that?** If necessary, allow the pupils to adjust their predictions.

Carry out the activity and compare their predictions with the actual distance covered. Ask: **How close was your prediction? How easy was it to predict how far your line would reach? Who had the most accurate prediction?** Discuss the length of lines drawn: **Who drew the shortest / longest line? Does anyone think that their line is half / double the length of this line?**

Enabling prompt

Provide opportunities for the pupils to experience using sand and electronic timers in timed activities, e.g. **I wonder if we can . . . tidy up the jigsaws** or **do 10 jumps** or **put our shoes on** or **build a tower** before the sand timer runs out. It is important to use different periods of time – 30 seconds / 1 minute / 2 minutes, etc. to help the pupils grasp the passage of time.

Extension activity

Challenge pupils to measure the lines drawn. Have available a range of materials, e.g. string / wool, pebbles, lollipop sticks, pine cones, stones, cubes, pencils, wooden blocks, etc. Ask: **What materials do you think will be best to use for the line you want to measure? Why do you think that? Which materials might be best to use if the line is very long? What would be good to use if the line was short?**

Let's check!

For the main problem, the pupils are being asked to guess how far their line will extend and then compare it to the actual line drawn in the period of time allocated. How close were their estimates?

For the extension activity, the pupils will choose materials to measure the length of the line. Can they use non-standard units effectively to measure the length of the line? Do they keep the unit constant whilst measuring, are the units touching end to end and do they use appropriate language to describe what they did? Certain materials may be more appropriate to measure the lines than others, e.g. lollipop sticks may be more appropriate for longer lines than pebbles.

Let's discuss!

Amman and Isla

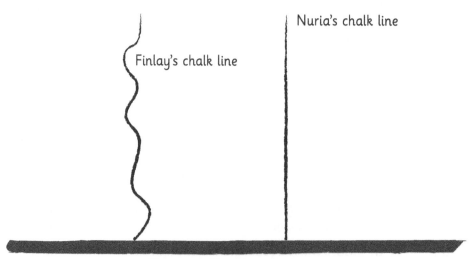

Look at Amman and Isla's chalk lines. **Who has drawn the longer line? Whose estimate was closest in length to the line that they drew? How close was your estimate?**

Finlay and Nuria

In the time given, Finlay and Nuria's lines ended at the same point in the playground. **Do you think that their lines are the same length? Can you tell me why you think they are the same length / different lengths? How could we check?**

Let's reflect!

This activity focuses on estimation in relation to measurement; the pupils are exploring how long and how far.

For the extension activity, it is important to reflect on measuring using non-standard units, where the measurements are expressed by counting the total number of units used. Do their explanations suggest that they understand that the measurement unit must remain constant? Do they appreciate that the units need to be placed correctly ensuring there are no gaps or overlaps?

Passage of time is a tricky concept for pupils to grasp. In this problem the pupils are drawing lines to help them visualise a period of time in relation to line length. Repeating this activity should result in their estimations becoming more accurate.

Teacher Guide Reference

- Chapter 6 – Time
- Chapter 7 – Measurement

How Many Are On The Bus?

Experiences and Outcomes

- ***MNU 0-03a** – I can use practical materials and can 'count on and back' to help me understand addition and subtraction, recording my ideas and solutions in different ways.*

Numeracy and Mathematical Skills

- **Interpret questions** – selects the relevant information; draws diagrams
- **Justify choice of strategy used** – shows and talks through their thinking
- **Uses mathematical vocabulary and notation** – uses developmentally appropriate mathematical vocabulary
- **Mental agility** – knowledge of number facts; manipulates numbers
- **Reason algebraically** – finds the unknown quantity

Resources

Interactive whiteboard images; selection of cardboard / plastic figures to act out the problem; pupils' choice of concrete materials if required; number lines (1–10 / 1–20); bead strings; ten frames, double ten-frames, bus-shaped counting mat. See Resource sheet – How Many Are On The Bus?

Before they start...

Pupils need to be able to count on and back within 10 for the Enabling prompt and within 15 for the Extension activity.

Let's go!

Read the problem to the pupils.

As the bus leaves the bus station, there are two passengers on the bus. At the first bus stop one passenger gets off and two passengers get on. At the second bus stop one passenger gets off and three get on. At the third stop four passengers get off and four get on. **I wonder how many people are on the bus now.**

Stop A	−1 + 2
Stop B	−1 + 3
Stop C	−4 + 4

This problem can be used many times. Use the interactive whiteboard images and alter the numbers of passengers entering and exiting the bus and/or the number of bus stops to create an appropriate problem for the pupils you are working with (see Extension activity). Encourage the pupils to think about what manipulatives / materials they might want to use to solve the problem.

I wonder how many passengers are on the bus now. How might we work this out? What do we know? What should we do first? What might we use to help us solve the problem? How do you think you are going to solve this?

Enabling prompt

Ensure that the pupils are familiar with the numbers getting on and off the bus. Check that they understand that the passengers getting on the bus equates to adding more passengers (counting on) and passengers getting off the bus equates to taking away some passengers (counting back).

It might be necessary to 'chunk' the problem for these pupils, i.e. by splitting up the whole route into separate stops. **How many passengers do you think are on the bus when it leaves the first bus stop?** Alternatively, consider having just one stop for the pupils to calculate and vary the numbers getting on and off. Pupils may benefit from role-playing the situation, or by using cardboard figures to represent the passengers.

Extension activity

Once the children are confident using one or two stops, increase the number of stops and include zero passengers getting off the bus at one of the stops. **How does this change the number of people on the bus?** Challenge pupils to represent the action of the problem using numerals and symbols. **When a person gets off the bus, are there more or fewer people left?** Encourage the children to use their own words to describe the activity at each bus stop. **How did you keep track of the numbers being added on and taken away?** Do any of the pupils see a quick way to calculate the number of passengers at each stop? How good are the pupils at keeping a running total? Do they use manipulatives? Do they make jottings on a whiteboard?

Stop A	−1 + 4
Stop B	−2 + 5
Stop C	−0 + 4
Stop D	−4 + 5

Let's check!

This activity can be carried out multiple times. The pupils will quickly grasp the requirements of the problem, become confident with the signage used and 'How many are on the bus?' can become an independent activity for the pupils to tackle with little or no adult support. It is important to observe how the pupils start the activity and what they do when carrying out the activity:

- What do they decide to use to represent the problem (fingers / manipulatives / drawing / number line)?
- Do they keep track using their fingers when the problem is being read or do they require to have the visuals of the bus stop figures to help with the count?
- Do they count all when they use their chosen manipulative, or do they count on?

Let's discuss!

This is how Isla, Amman, Finlay and Nuria solved the problem. I wonder if you did something similar.

Isla decided to use her fingers to count the people getting on and off the bus. She found she had to concentrate quite hard to keep track. **Did you use your fingers to keep track of your counting? Do you think this is a good way to work out the problem? Why / why not?**

Amman chose to use counters; he placed them on and took them off the bus picture to show the passengers on the bus after each stop. He worked out that the answer was five. **Did you use counters to solve the problem? What did you do with your counters? Show me.**

Finlay used the bus template. He decided to use matchsticks to model the passengers getting on and off the bus.

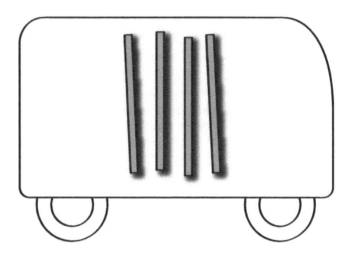

Do you think Finlay made a good choice using sticks? How did you solve the problem?

Nuria decided to use a number line. She looked at the numbers of passengers getting on and off the bus. She estimated that the number of people on the bus might be more than 10, so she chose a 1–20 number line. She decided to draw her 'counting on' jumps at the top to show the people getting on the bus and her 'counting back' jumps at the bottom to show the people getting off. She worked out that 11 people were on the bus altogether but 6 people got off so that left 5 people on the bus at the end of the journey.

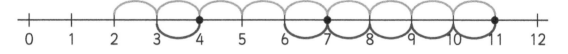

Do you think Nuria's method was a good one? Why did she start counting on from two and not zero? Did you use a number line? How did you work it out? What did you do?

Let's reflect!

Isla chose to solve the problem by counting on her fingers but had difficulty keeping track of her count. Ask pupils, **Why do you think this happened? What would you say to Isla to help her?**

Amman initially tried to draw a bus shape to put his counters on; he found this hard and it hindered his problem solving. Providing him with a bus-shaped template (available for use in Resources) reduced his cognitive load and he was able to move on with the problem. Amman understood the signage used and said it was take away and add on. Amman could progress to using more bus stop stages and extending the number range.

Finlay used matchsticks to represent the people and managed to solve the problem correctly. He could be challenged in a similar way to Amman.

It is important that the pupils see different ways of solving the problems using different manipulatives; they should feel confident to change the way they are modelling the problem and try using a different resource if their choice is proving inefficient.

Nuria showed evidence of linking her learning and estimated that the number of passengers on the bus may go above 10; she rightly chose the 1–20 line. She showed deep thinking and creatively found a useful solution by marking lines above and below the number line. She could be challenged to write a number sentence to mirror her thinking and encouraged to reflect on the impact that the same number of passengers getting on and off the bus (3rd stop) has on the total.

Teacher Guide Reference

- Chapter 2 – Number and Number Processes – Number Recognition and Place Value
- Chapter 3 – Number and Number Processes – Four Operations

Number Lines

Experiences and Outcomes

- *MNU 0-02a – I have explored numbers, understand they represent quantities, and I can use them to count, create sequences and describe order.*

Numeracy and Mathematical Skills

- **Select and communicate processes and solutions** – shares thinking
- **Justify choice of strategy used** – shows and talks through their thinking
- **Use mathematical vocabulary and notation** – uses developmentally appropriate mathematical vocabulary

Resources

- Wonder jar

- An empty number track (for the Enabling prompt, available as a downloadable resource)
- A length of string and pegs to represent a washing line (for the Enabling prompt)

- Strip of paper with empty number line marked on it to place in the wonder jar (for the Extension activity)
- Small number tiles (0–30)
- Pens / pencils
- A long strip of paper marked with an empty number line per pair of pupils
- Optional completed number lines for self-checking

0				
1	2	3	4	5
6	7	8	9	10
11	12	13	14	15
16	17	18	19	20
21	22	23	24	25
26	27	28	29	30

Before they start...

Pupils need to be able to recognise, identify and order numbers within the range 0–30 (up to 10/20 for enabling prompt).

Let's go!

I wonder what our challenge is today in the wonder jar? Open the jar and lift out the number line that has been rolled up. Unroll to reveal an empty number line. Ask, **What do you think this might be? Have you seen something like this before?** Once you are satisfied that the pupils have had time to share their ideas, tip out the number tiles from the bottom of the jar. Tell the pupils that the numbers have fallen off the number line and you need their help to put them back on. Ask, **What should we do first? What do you think we should look for? Is there anything else we need?**

Enabling prompt

To help pupils who are struggling to get started with the problem:

- reduce the number range, e.g. provide the number tiles 1–10 (or an appropriate range for the pupils involved) to place on a number track, washing line, etc.
- provide a partially completed number track / number line. Ask the pupils – **What numbers can you see on the number track? Which numbers might be missing? What comes after 1, before 7, etc?**

0	1		4		7	

Extension activity

To provide challenge:

- Ensure that not all of the numbers are in the jar. Which numbers are missing?
- Include some numbers outwith the range 0–30, e.g. 31 and 32.
- Place in the jar number tiles starting from, e.g. 5: some pupils may need practice in starting their count from a number other than 1.

Which numbers are missing? Can you find a number tile that would fit into this space? Which number tiles don't fit?

Let's check!

There is one solution to this problem (a completed number line with the numbers in the correct order) but the way in which the pupils approach it can indicate the depth of their understanding of number order, numeral identification and numeral recognition. Listen and look carefully.

Let's discuss!

Enabling prompt

Amman and Nuria thought that the empty number line looked like the class washing line. They decided they would look at all the number tiles they had been given. They wondered what might go first. **What number did you put first?** Amman wondered if they should start with a zero but couldn't find one in the pile. Nuria said they should start with the number 1 because you can't put zero on a number line. **Do you think Nuria was correct?** Together they compared the other numbers using bigger than and smaller than; they thought that they had two number 6 tiles, but Amman turned his round and found it was a 9. Amman counted out loud and Nuria found the numbers and placed them on the line. **What did you do? How did you start? Did you do this? Did you manage to put all the numbers in the correct order?**

Extension activity

Finlay and Isla sorted their numbers into piles: single digits, tens and twenties. **How many piles do you think they had? How many number tiles do you think there were in each of the piles? How did you sort the tiles?** They found that they had three piles of number tiles and one number tile left over. **Can you think what number tile might be left over?** They completed 0–11. Finlay placed the next tile, he used 21 but Isla said he was wrong. **What do you think? What number tile should he have used?** Isla counted out the next three tiles. As she placed them, she said, '30, 40, 50'. Finlay said, '13, 14, 15'. **Who do you think is correct?** Finlay and Isla completed their number line. They realised that the leftover number was 30 and placed it at the end.

Let's reflect!

This is a versatile task that can be useful in assessing what the children understand about number order.

Some things to look and listen for:

- Are the pupils seeing or identifying 12 as 21, 13 as 31, etc? Practice in numeral recognition and numeral identification with flashcards may be useful.
- Are the pupils saying thir**ty** instead of thir**teen?** If so, make sure you are saying the words clearly enough – emphasise the *'teen'* in thirteen and the *'ty'* in thirty.
- A sound knowledge of number before, number after and number in between.

When using an empty number line for counting, it is important to emphasise that 'jumps' are counted, not numerals. Zero should therefore be included; however, this idea may be too abstract for some.

Pupils who are not yet secure in their understanding of the Counting Principles may confuse the number word sequence (0, 1, 2, 3, 4 . . .) with counting to determine 'how many' (1, 2, 3, 4 . . .). These pupils are not yet ready to be introduced to the empty number lines and should instead be given a number track or washing line starting at 1, where concrete items (boxes and 'washing') can be counted.

Teacher Guide Reference

- Chapter 2 – Number and Number Processes – Number Recognition and Place Value

Juggling Clowns

Experiences and Outcomes

- *MNU 0-02a* – *I have explored numbers, understanding that they represent quantities, and I can use them to count, create sequences and describe order.*
- *MNU 0-03a* – *I use practical materials and can 'count on and back' to help me understand addition and subtraction, recording my ideas and solutions in different ways.*

Numeracy and Mathematical Skills

- **Justify choice of strategy used** – shows and talks through their thinking
- **Use mathematical vocabulary and notation** – uses developmentally appropriate mathematical vocabulary
- **Mental agility** – knowledge of number facts

Resources

- Resource sheet – Juggling Clowns
- Image of juggling clown
- Juggling ball cards (available in downloads)
- Numeral cards (rectangular) for the total (available in downloads 0–20)
- Concrete materials – bead strings, counters, buttons, empty ten frames etc.

Before they start...

The pupils should be able to:

- add two or three single-digit numbers (with or without concrete materials) to find a total
- identify and recognise numbers within 20

Let's go!

The clown is juggling three balls. I wonder what numbers are on them. Turn over three juggling ball cards. What can you see? How could you find the total? Pupils should select a rectangular number card that matches their answer and place it on the clown's jumper.

Enabling prompt

If the pupils are unable to subitise, encourage them to count the dots or lay out concrete materials. If they can subitise then use regular dot pattern cards from one to three. Encourage the pupils to pick two or three cards at random and arrange them above the clown. **What numbers are on the clown's juggling balls? We can find the total by adding the numbers together. How will you do it?** After using dot cards, progress to numeral cards. Extend the number range as appropriate.

Extension activity

Place a number card in the clown's box. **The clown is juggling three balls. I wonder what numbers could be on the balls to make this total. Can you find three juggling balls that will make this total?**

Introduce a fourth juggling ball to increase the level of challenge further.

Let's check!

Solutions 0–6 Juggling balls.

6 + 5 + 4 = 15	6 + 4 + 3 = 13	6 + 3 + 2 = 11	6 + 2 + 1 = 9	6 + 1 + 0 = 7
6 + 5 + 3 = 14	6 + 4 + 2 = 12	6 + 3 + 1 = 10	6 + 2 + 0 = 8	
6 + 5 + 2 = 13	6 + 4 + 1 = 11	6 + 3 + 0 = 9		
6 + 5 + 1 = 12	6 + 4 + 0 = 10			
6 + 5 + 0 = 11				
5 + 4 + 3 = 12	5 + 3 + 2 = 10	5 + 2 + 1 = 8	5 + 1 + 0 = 6	
5 + 4 + 2 = 11	5 + 3 + 1 = 9	5 + 2 + 0 = 7		
5 + 4 + 1 = 10	5 + 3 + 0 = 8			
5 + 4 + 0 = 9				
4 + 3 + 2 = 9	4 + 2 + 1 = 7	4 + 1 + 0 = 5		
4 + 3 + 1 = 8	4 + 2 + 0 = 6			
4 + 3 + 0 = 7				
3 + 2 + 1 = 6	3 + 1 + 0 = 4			
3 + 2 + 0 = 5				
2 + 1 + 0 = 3				

Let's discuss!

Amman and Isla had a set of irregular dot pattern cards. Isla turned over three cards.

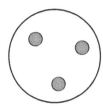

Amman knew that this was 3, without counting.

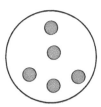

Isla didn't recognise this pattern. However, she saw two dots and three dots and knew that made five dots. **What do you see when you look at this dot pattern? Tell me how you see the dots.**

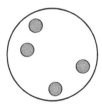

Amman saw two lots of two and knew that made four.

Isla started counting in ones to find out how many dots there were altogether, but Amman said he could find the total without having to count. **What do you think Amman did? Can you think of a different way to find how many dots there are together? Which ball would you start with? Why?**

Extension task

Finlay and Nuria picked a number card: $\boxed{12}$

They wondered what three juggling ball cards they could use to make this number. Nuria said 6 + 6 + 0 would be a solution. **Could Finlay and Nuria have scored 6 + 6 + 0? Why do you think that? What other ways are there to make 12?** [The solution suggested by Finlay and Nuria is not possible as they only have **one** set of juggling ball cards, 0–6. To introduce more solutions, add another set of juggling ball cards.]

Let's reflect!

This task provides an opportunity to discuss the conservation of number in context; the scores on the juggling balls may appear in different orders but give the same total.

Listen carefully when the pupils are explaining how they added together the numbers on the juggling balls. Do they add in the order they see them? Do they put the 'big' number first? Do they look for combinations that they know?

If the pupils are using the Extension task, ask, **How can we be sure that we have found all the possibilities?** Further investigation may involve asking questions such as **What is the biggest number that can made using three juggling balls? What is the smallest number that can be made using three juggling balls? Tell me how you worked this out.**

Teacher Guide Reference

- Chapter 2 – Number and Number Processes – Number Recognition and Place Value
- Chapter 3 – Number and Number Processes – Four Operations

Parking Cars

Experiences and Outcomes

- ***MNU 0-02a*** – *I have explored numbers, understanding that they represent quantities, and I can use them to count, create sequences and describe order.*

- ***MNU 0-03a*** – *I use practical materials and can 'count on and back' to help me understand addition and subtraction, recording my ideas and solutions in different ways.* (see Extension activity)

Numeracy and Mathematical Skills

- **Select and communicate processes and solutions** – shares thinking
- **Justify choice of strategy used** – shows and talks through their thinking
- **Uses mathematical vocabulary and notation** – uses developmentally appropriate mathematical vocabulary

Resources

- Numbered cars (see Resource sheet - Parking Cars)

- Blank car template (downloadable)

- Digit cards (0–22), dot pattern cards, 5 frames, 10 frames
- Five-wise ten frames

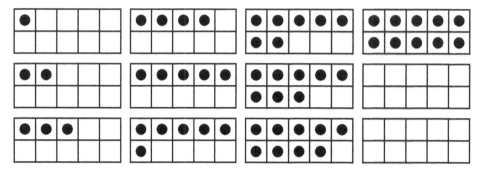

- Pair-wise ten frames

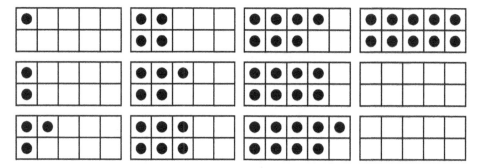

- Wipe-clean markers – for use if cars are laminated
- Number tracks (for self-checking)

Before they start...

Pupils should have experience of:

- recalling number word sequences forwards and backwards
- recognising the number before, after, in-between
- using ordinal numbers (e.g. car number 3 is first in the queue)
- identifying and recognising numerals from 0–20
- finding the difference between two numbers

Let's go!

Distribute a set of numbered cars to individuals or pairs (vary the numbers on the cars, depending on the experience of the pupils). Ask, **What numbers can you see? Which is the biggest number? Which is the smallest number? What can you tell me about this car / number?**

Tell the pupils that you would like them to park the cars in order. Observe what they do. Do they use all of the cars or only some of them? Can they explain their order?

Which number did you start with? Which number car did you park last? Why did you put car number * there? Can you park the cars in a different order? How did you park them?

Pupils can be challenged to:

- sequence consecutive numbers from smallest to largest or from largest to smallest
- order a random selection of numbers, e.g. 21, 6, 9, 10, 19, 20, 4, 12, 17, 5, 7
- park only the even or odd numbered cars

Enabling prompt

Provide opportunities for the pupils to experience ordering numbers within the range 0–10. Use resources that the pupils are familiar with, e.g. number tiles, numbered buttons, numbered building bricks, etc. Include cars marked with dot patterns / 5 frames / 10 frames for pupils who have difficulty recognising numerals. **What numbers can you see? Can you find the number . . .? Which number will go after 3? Can you put these numbers in order? Which number goes between . . . and . . .?**

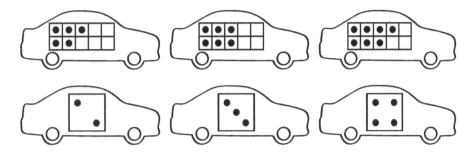

Extension activity

- Provide a selection of numbered cars for the pupils to park. Their challenge is to ensure that the difference between each pair of parked cars is 2; for example 8, 10, 12, 14 . . .
- Increase the number range.
- Extend the range of cars offered to include numbers greater than 30.

Let's check!

A range of solutions will be possible, depending on the car templates that the pupils choose or are provided with. The advantage of this task is that all pupils can use the same resource but will solve the problem at their own level of understanding.

Let's discuss!

Nuria and Finlay looked at the cars they had been given.

They arranged them in a row and checked to see that they could identify all of the numbers. They decided to put the numbers in order from smallest to largest but were unsure which number to start with. **Can you think what they might be finding tricky? How could we help them get started? Which number do you think will come first?**

Isla was given these cars to park in order.

She decided to order the numbers from largest to smallest. **Which car do you think she started with?** Isla used her number track to help her decide which car would come first. She wasn't sure whether it was this one, or this one (hold up cars 21 and 12). **Which car do you think Isla should choose? Can you think why she found this tricky? Which car number will come last?**

Amman was given a bundle of cars to park. He had to make sure that the difference between each parked car was 2. He thought that if he found the smallest number on any of the cars and counted on 2 that should give him the number of the next car. **Do you think Amman was right? Can you tell me why you think that?**

Let's reflect!

While pupils are solving the problem, look for them showing confidence in, and awareness of, number order and the counting sequence; listen carefully to what they are saying. Some pupils may struggle to order numbers if the numeral 1 isn't present. It is important that they realise that they need not start at one. They should be presented with opportunities to count orally from a range of numbers. Check that numbers are not being misplaced, e.g. 12 / 21 and that the pupils have opportunities to cross decades. Listen carefully to the enunciation – are they saying 'thirteen' or 'thirty'? Numeral identification and recognition are very important in place value understanding; pupils should experience a wide range of activities to challenge and extend their learning.

Teacher Guide Reference

- Chapter 2 – Number and Number Processes – Number Recognition and Place Value
- Chapter 3 – Number and Number Processes – Four Operations

It's Teddy's Birthday!

Experiences and Outcomes

- *MNU 0-02a – I have explored numbers, understanding that they represent quantities, and I can use them to count, create sequences and describe order.*
- *MNU 0-03a – I use practical materials and can 'count on and back' to help me understand addition and subtraction, recording my ideas and solutions in different ways.*

Numeracy and Mathematical Skills

- **Interpret questions** – selects the relevant information
- **Select and communicate processes and solutions** – explains choice of process; shares thinking
- **Justify choice of strategy used** – explains their strategy
- **Use mathematical vocabulary and notation** – uses developmentally appropriate mathematical vocabulary
- **Mental agility** – knowledge of number facts

Resources

- A teddy bear or a picture of a teddy bear.
- Access to a range of manipulatives, e.g. interlocking cubes, bead strings, counters, five/ten frames
- Paper and pencil
- Birthday cake pictures (see Resource sheet - It's Teddy's Birthday)

Before they start...

Pupils should be able to

- count a given number of objects up to 15
- recognise when addition is required
- use concrete materials (e.g. interlocking cubes, bead string, counters, five/ten frames) to find a total

Let's go!

Read the following problem to the pupils:
It is Teddy's birthday today. He is five. Each year Teddy is given the same number of presents as his age. I wonder how many birthday presents Teddy has been given since he was born?

What do you think we will need to do to solve this problem? I wonder what we should do first.

What do we know? How many birthday presents did Teddy get today? I wonder how many presents Teddy got last year? How do you know?

Enabling prompt

For learners who are struggling to make a start to the problem, reduce Teddy's age to three. If they are still struggling, provide the pupils with the birthday cake images (Resource sheet – It's Teddy's Birthday) that have the appropriate number of candles already attached.

Extension activity

Re-read the problem to the pupils but this time introduce Big Ted! Big Ted is older than Teddy (choose an appropriate age for Big Ted to suit the degree of challenge required). Use the questions in Let's go to ensure that all pupils have grasped the important information. **How will you solve the problem? Can you think what you might do first? What do you know already? How might this help you? Can you estimate how many parcels Big Ted might have got altogether? How will you check your estimate?** Listen to the discussions that each pair of children are having, noting the use of mathematical vocabulary.

Let's check!

A Teddy who is three would have received 6 presents in total 1 + 2 + 3

A Teddy who is five would have received 15 presents in total 1 + 2 + 3 + 4 + 5

Let's discuss!

Amman and Finlay worked together. They used Amman's teddy, rather than using pictures. **Did you use one teddy or did you use pictures?** They agreed that using cubes for parcels might help them work out how many presents Teddy had been given. They collected a pile of cubes and counted them out; 1 present for when Teddy was one, 2 presents for when he was two, etc. They ended up with a big pile of cubes. **Do you think having a pile of cubes is a good idea?** Finlay tried to count all of the cubes that were sitting in front of Teddy. He wasn't sure if he had counted them all once. Amman suggested that it would be easier to count the cubes if they put them in a line. **Do you think that was a good idea? Why do you think that? How else could he count the presents so as not to get mixed up?** Finlay touch counted the cubes and got 15 as his answer. Amman checked it and he said 15 was the answer he got too. **Is this the answer you got? Is this how you worked it out? Did you find a different way to work it out?**

Nuria and Isla worked together to solve the problem. Teddy was five today. Isla said that he would have 5 presents. Last year he would have had 4. She and Nuria were good at counting back and knew that the other years he would have been given 3, 2 and 1 presents. They wrote out the addition problem they thought they'd need to do to count all the presents: 1 + 2 + 3 + 4 + 5. **Do you think this will give them how many presents Teddy had altogether? Why do you think that? The girls added together 1 and 2 to make 3; then 3 and 4 to make 7. They know that 7 and 3 = 10.** They crossed out the numbers 1, 2, 3 and 4 because they had already been added and noticed they hadn't used 5 yet. They counted on 5 from 10 to get an answer of 15. **Do you think this was a good strategy? How did you do it?**

Let's reflect!

Observe how pairs of pupils start the task, the manipulatives they choose, the mathematical conversations they have, the methods that they use and the solutions they arrive at. The way the pupils approach the problem is indicative of their developmental stage.

Amman and Finlay are confident counters. They counted out the cubes to reflect the number of presents into one pile or line and used one-to-one correspondence to find the total. Observing the solutions that other children share may encourage them to try other approaches when tackling similar problems. It is important to ensure that all pupils are secure with all of the Counting Principles.

Explore which might be the best manipulative to use for this type of task with the pupils. **Would using empty ten frames and counters have made counting easier? Why?** Encourage the pupils to justify which they think might be best. Although the pupils may have a favourite 'go to' manipulative, they should be given opportunities to experience a range so that they can begin to develop an appreciation of the merits of each.

Nuria and Isla have developed one-to-one correspondence and can confidently count forwards and backwards. They recognised that the problem could be solved using addition and were able to add mentally using their knowledge of counting on, doubles, near doubles, partitioning and known facts to find their solution. They were able to explain what strategies they used. Next steps for pupils like Isla and Nuria could be to explore a wider range of problem types, e.g. odd number plus an odd number makes an even number, to challenge their thinking.

Teacher Guide Reference

- Chapter 2 – Number and Number Processes – Number Recognition and Place Value
- Chapter 3 – Number and Number Processes – Four Operations

Towers Using 10

Experiences and Outcomes

- *MNU 0-02a – I have explored numbers, understanding that they represent quantities, and I can use them to count, create sequences and describe order.*
- *MNU 0-03a – I use practical materials and can 'count on and back' to help me understand addition and subtraction, recording my ideas and solutions in different ways.*

Numeracy and Mathematical Skills

- **Select and communicate processes and solutions** – shares thinking
- **Justify choice of strategy used** – shows and talks through their thinking
- **Use mathematical vocabulary and notation** – uses developmentally appropriate mathematical vocabulary

Resources

- 10 building blocks
- Cubes (see Enabling prompt)
- Part-whole mats (see Extension activity; Resource sheet – Towers Using 10)

Before they start...

The pupils should have had opportunities, through practical experiences, to explore ways of making 10.

Let's go!

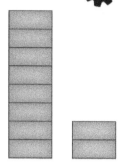

Can you build a castle using 10 blocks? Your castle can have more than one tower. Each tower must have a different number of blocks.

Here is a castle built using two towers. I wonder what other castles we could build using ten blocks. Allow time for the pupils to think about what they might do before asking them to collect 10 building blocks.

Enabling prompt

Check that the pupils have collected the correct number of cubes / bricks / blocks required for the problem.

If necessary, reduce the number of blocks and provide some pictorial examples for pupils to copy. For example, **How many bricks are in each tower? How many bricks will you need altogether to make these towers? What is the same or different about the towers you have made?**

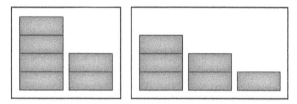

Extension activity

Provide the pupils with a selection of laminated part-whole mats, wipe-clean markers and cubes.

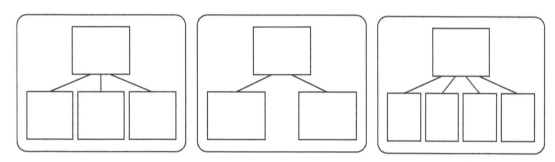

Can they illustrate their findings using the part-whole mats? How many different ways can they find to make 10? They may choose to display their findings by using cubes / counters, writing numbers or a mixture of both. For example:

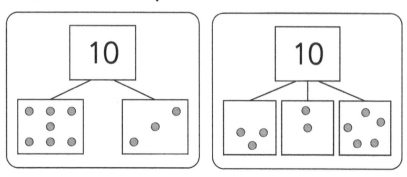

Let's check!

Solutions to the problem:

One tower – 10 bricks

Two towers – 9 & 1, 8 & 2, 7 & 3, 6 & 4 bricks

Three towers – 7 & 2 & 1, 6 & 3 & 1, 5 & 3 & 2, 5 & 4 & 1 bricks

Four towers – 4 & 3 & 2 & 1 bricks

Let's discuss!

Finlay and Amman took 10 bricks. They thought they would start by building a tower of 10. Amman suggested that they could split their tower into two smaller towers. He built a tower with 6 bricks and a tower with 4 bricks. Finlay was sitting opposite Amman and said he could see a tower of 4 bricks and a tower of 6 bricks. **How can they both be correct?**

Nuria and Isla wanted to build a castle with three towers. They thought it would look best if they had a tall tower and two shorter towers. They took 10 bricks. Isla built a tower of 6. She could see that there were 4 bricks left but remembered that the towers couldn't be the same height. **Can you think what towers Isla built to solve the problem?** The girls wondered if there were any other three-towered castles that could be made. **Who has a different solution to Isla and Nuria? Can you share your solution with us? What number of bricks do you have in your towers?**

Let's reflect!

This is a hands-on practical task; the pupils are manipulating the materials and have the opportunity to act out the problem. The end-product – the castle and towers – represents a great window to their thinking. Ask: **Do you think we have found all the possible solutions? How do you know?** The problem provides opportunities to explore the commutative nature of addition and how the same quantity can be partitioned in different ways without affecting the total.

Teacher Guide Reference

- Chapter 2 – Number and Number Processes – Number Recognition and Place Value
- Chapter 3 – Number and Number Processes – Four Operations

Which Number Goes Where?

Experiences and Outcomes

- *MNU 0-02a – I have explored numbers, understanding that they represent quantities, and I can use them to count, create sequences and describe order.*

Numeracy and Mathematical Skills

- **Justify choice of strategy used** – shows and talks through their thinking
- **Use mathematical vocabulary and notation** – uses developmentally appropriate mathematical vocabulary
- **Mental agility** – manipulates numbers

Resources

- Interactive whiteboard images
- Printed empty 'templates' of the problem (see Resource sheet – Which Number Goes Where?)
- Whiteboards and pens
- Number tracks
- Cubes (See Enabling prompt)

Before they start...

The pupils should be able to:

- order numbers (orally or using labelled concrete objects, e.g. digit cards, number buttons, post-its, etc.)
- describe the relationship between single-digit numbers using the words bigger and smaller

Let's go!

Show this puzzle to the pupils. Ask, **What can you see? What do you think you are being asked to do?** Check that the pupils recognise the numbers and that they understand the direction of the arrows. Allow them time to think and talk about the problem with a partner.

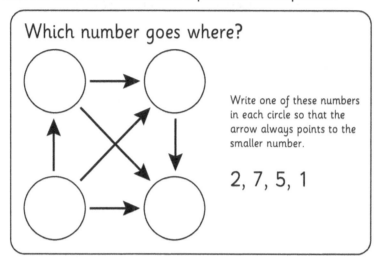

Which number goes where?

Write one of these numbers in each circle so that the arrow always points to the smaller number.

2, 7, 5, 1

Enabling prompt

Check that pupils understand the meaning of bigger and smaller.

Provide the puzzle in a linear format and ask the pupils to arrange the numerals from smallest to biggest.

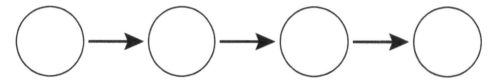

Which circle should we start with? Tell me why you think that. Which number shall we put in the first circle? Which number should we put in the next circle? Model appropriate 'ordering' vocabulary, e.g. largest / larger; biggest / bigger; smaller / smallest; first / next / before / after, etc.

OR

Provide cubes that the pupils can use to build towers that represent the numbers in the problem. Ask: **Which tower has most cubes? Can you put the towers of cubes in order from most cubes to least cubes? How many cubes are there in each tower? How can this help us solve the puzzle?**

Pupils can then be challenged to display their cube towers in the linear diagram and, if appropriate, in the original diagram.

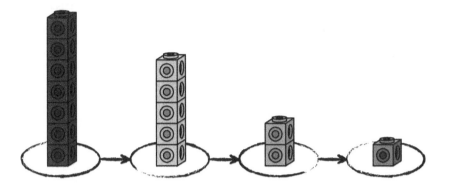

OR

Provide the pupils with counters and paper plates. The counters can be arranged in regular or irregular dot patterns, e.g.

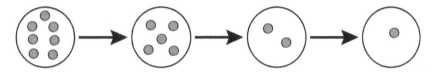

Extension activity

Challenge the pupils to solve this puzzle.

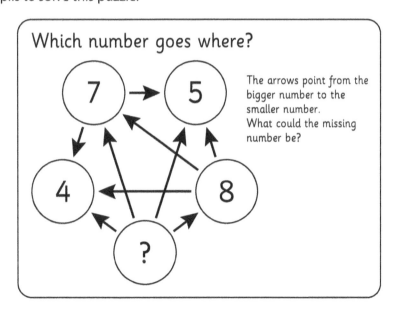

Which number goes where?

The arrows point from the bigger number to the smaller number.
What could the missing number be?

OR

Provide an empty template (see downloadable resource). Challenge the pupils to work in pairs to create their own puzzles that satisfy the rule, i.e. arrows always point to a smaller number.

Let's check!

The solutions will vary depending on the actual numbers used and the layout of the problem; however, the underlying principle of the arrow pointing from a number towards a smaller number remains constant.

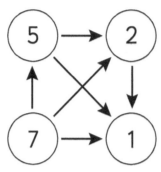

Let's discuss!

Finlay and Nuria were working together; Nuria wrote the numbers down in the order: 7, 5, 2, 1. **How might this help them?** Finlay and Nuria looked at the arrows and saw that all the arrows were pointing away from one of the circles. They decided to put the 7 in that circle. **Why do you think they did that? How might that help them? Where should the number 5 go?** Nuria and Finlay quickly solved the rest of the puzzle. **Where do you think they put the 2 and the 1?**

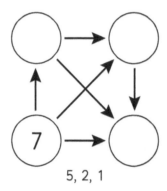

5, 2, 1

Which number did you start with? Why did you pick that number? Where did you put it? Why did you decide to put it there?

Isla and Amman tried this puzzle.

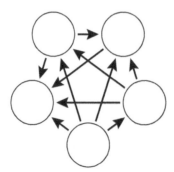

They chose five numbers. Amman suggested rolling a dice and using the first five numbers rolled but Nuria said they might get two numbers the same and all of the numbers had to be different. **Is Nuria correct? Do all the numbers have to be different?**

They took turns at thinking of numbers and chose 4, 12, 9, 11 and 6 to fit onto the template.

Amman said they had to look for a circle with all the arrows pointing away from it so they could get started. **Can you tell me why Amman wants to look for a circle with all the arrows pointing away from it?** Isla said they should put their biggest number in it. **Which number would you put in it?** Isla spotted where the smallest number would go. **Which number is the smallest? Where do you think it should go? Tell me why you think that.** Amman and Isla finished the puzzle. **Where do you think they put the other numbers?**

Let's reflect!

This problem provides an opportunity for the pupils to apply their knowledge of number order; they have to work systematically and use reasoning to solve it. Listen carefully for the use of appropriate mathematical vocabulary. For pupils who are not confident in writing numbers, provide numbered counters or digit cards.

Teacher Guide Reference

- Chapter 2 – Number and Number Processes – Number Recognition and Place Value

Stone Match

Experiences and Outcomes

- **MTH 0-16a** – I enjoy investigating objects and shapes and can sort, describe and be creative with them.
- **MTH 0-17a** – In movement, games and using technology I can use simple directions and describe positions.

Numeracy and Mathematical Skills

- **Justify choice of strategy used** – shows and talks through their thinking
- **Uses mathematical vocabulary and notation** – uses developmentally appropriate mathematical vocabulary

Resources

- Selection of stones in a basket or box (always include a few extras that haven't been drawn round)

- Stone Shape Sheet (draw around a number of the stones ensuring that the outlines require the pupils to manipulate the stones to get a fit)

Before they start...

Pupils should have experience of manipulating shapes and objects, e.g. using a shape sorter where they need to flip or rotate shapes to fit a space. Inset puzzles, nesting toys, Russian dolls and jigsaws are also useful.

Let's go!

Show the pupils a picture of the activity on an interactive whiteboard.

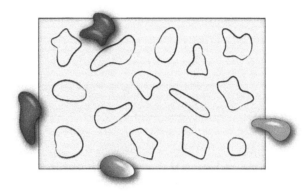

Provide pairs of pupils with a basket of stones and a sheet similar to the image shown here. **Tell me about the stones that you can see.** Talk about the shapes on the sheet. **What do you notice? What do you wonder?** Ask: **Can you solve this puzzle by finding a stone to match each outline?**

Enabling prompt

Provide outlines of regular 2D shapes and ask the pupils to match the correct shape to each. **Does it fit? How can we make it fit?**

Some 2D shapes should require rotation or flipping.

Extension activity

Provide the pupils with tangram puzzles to solve. Can they fit the shape pieces together to make the picture?

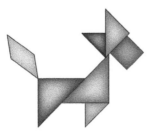

Let's check!

This problem provides the pupils with the opportunity to develop their visual perception and spatial reasoning skills as they manipulate the objects. Look for pupils rotating and flipping the stones to get a best fit and using descriptive language in relation to the shape, size and position of the stone.

Let's discuss!

Finlay and Nuria looked at the shapes on the sheet. Nuria said she saw a sausage shape and a heart shape and Finlay saw a square shape and an oval shape. **Which outlines do you think they are thinking about? What makes you say that?** They laid out all the stones and looked for ones that they thought might match the outlines. When Finlay found the sausage-shaped stone it looked upside down. **What do you think Finlay had to do to get the stone to fit? What would help Finlay and Nuria match some of the other stones? What should they look for?**

Amman and Isla looked at the shape outlines. Amman said that the outlines were different sizes: six big stone shapes and seven smaller stone shapes. **How do you think Amman and Isla will sort their stones? How would you sort the stones? What might they need to do find a matching outline?**

Let's reflect!

The children managed to find the appropriate stones to fit the outlines. They were able to transfer previous learning in measure, shape, position and movement to this context and used appropriate vocabulary. Encourage the pupils to describe the changes in how the stone or shape looks when it is rotated and to look at objects from different perspectives.

Teacher Guide Reference

- Chapter 9 – 2D Shapes and 3D Objects
- Chapter 10 – Angle, Symmetry and Transformation

Stacking It Up!

Experiences and Outcomes

- *MNU 0-02a* – *I have explored numbers, understanding that they represent quantities, and I can use them to count, create sequences and describe order.*
- *MNU 0-11a* – *I have experimented with everyday items as units of measure to investigate and compare sizes and amounts in my environment, sharing my findings with others.*
- MTH 0-16a – *I enjoy investigating objects and shapes and can sort, describe and be creative with them.*
- *MNU 0-20b* – *I can match objects, and can sort using my own and others' criteria, sharing my ideas with others.*

Numeracy and Mathematical Skills

- **Select and communicate processes and solutions** – explains choice of process, shares thinking
- **Justify choice of strategy used** – shows and talks through their thinking
- **Link mathematical concepts** – transfers learning in one area to another
- **Uses mathematical vocabulary and notation** – uses mathematical vocabulary appropriate to their stage of development

Resources

- Stones of varying sizes and shapes (see Enabling prompt)
- Sea glass of varying size and shape (see Extension activity)
- Other materials that you have in your setting that the children could use to stack, e.g. junk modelling materials, loose parts (see Extension activity)

Before they start...

Pupils need to have experience of

- building with 3D objects
- sorting items using their own or a given criteria

Let's go!

Amman saw some stone stacks on a visit to the beach. He decided to build a stack to show Nuria what he had seen.

Show, and talk about, the picture with the pupils. **Does this look like anything you've seen before?** Explain to them that Amman wanted to make a tower as tall as this one, or an even taller one! Ask the children, **How many stones do you think there are? How many stones would Amman need?**

Has anyone seen a stone stack on a visit to the beach, or made a stone stack? Was it easy to build? I wonder if we could build a stone stack. What might we do to get started? Have we done anything like this before? Let's estimate how many stones we might be able to balance in our stack.

Encourage the children to have a go at stack building and share their thinking as they build. Note how they approach the task, listen carefully to what they say and challenge them to try to make a taller stack. Look for perseverance but also for children who are not afraid to start again and use a new approach.

Enabling prompt

For pupils who are finding stacking stones tricky, ensure that there are lots of flat stones available to choose from. **Can you remember when we built towers with our wooden blocks? Which shapes of blocks made the steadiest towers? Which stones do you think might make the best tower?** A nice starting activity could be to sort the stones into flat and not flat. The pupils may suggest or want to explore other materials to build stacks with. Ask questions such as: **I wonder how we could make your tower taller? I wonder how many stones you have used already?** When talking about the stones / sea glass / loose parts materials, model and encourage the use of **mathematical language**, e.g. edge / flat / curved, **positional language**, e.g. on top of / underneath and **informal language**, e.g. hard / rough / smooth. This will encourage the children to use it in their play and help them describe what they did.

Extension activity

The approach to this activity is essentially the same as the stone stacks; however, the nature of the sea glass adds some complications; it is light, has a smooth texture and is often irregular in shape. (If you do not have sea glass, then use another resource that can stack.)

Nuria loved the idea of building stacks but wanted to build her own with something prettier. She decided to use sea glass. **Can you tell me anything about the sea glass that is the same as / different to the stones? Which do you think will be easiest to build with? Why?**

Which pieces of glass might make a good base? Which shape is best to build with?

If you do not have access to sea glass, consider using junk modelling or loose parts materials; consider also using the stones that were sorted into the 'not flat' category.

Let's check!

Remind the pupils of the photograph Amman took of a stone stack. **When you built your stack, what did you find tricky? Did you find it quite easy? What advice would you give to someone to help them build a stack?**

Amman had fewer layers, but his stack was taller than Nuria's. Can you think why? Does tallness relate to the number of layers?

Observe and note whether the pupils can:

- Select and sort appropriate stones / sea glass according to size, shape and texture.
- Count the number of layers in their stack.
- Use appropriate vocabulary, e.g. taller than, shorter than, smooth, rough, shiny.
- Arrange the stones / sea glass in order to build the tallest, most stable tower they can.
- Talk about similarities to building with other materials and in different contexts, e.g. when using junk modelling, some shapes make a better base.
- Justify their choices.

Let's discuss!

Here are Amman, Isla, Nuria and Finlay's solutions to the problem. I wonder if anyone built stacks like theirs.

Amman chose some stones; he was keen to get started. He looked for some flattish stones but didn't think about the sizes.

He managed to build a stack that was five stones high. His top stone was curved on one side and he found it quite hard to put any other stones on top. He knew that five is a lot less than fifteen! **Why does Amman find it hard to place another stone on his stack? I wonder if we can suggest ways to help Amman? I wonder how many more stones Amman might need to make a stack of fifteen? Can you show me how you worked that out?**

Amman decided to knock his tower down and try again.

This time he managed to build a stack six stones high. He looked more carefully at the shape of the stones but not at the sizes. **What do you think Amman needed to do first? Did you look at the size and shape of your stones? What did you do?**

Nuria started by sorting her sea glass into flat and not flat. She looked closely at the flat pieces and put them in order of size. She knew that the largest flat piece would make a good base. Nuria found the pieces of glass quite smooth and found that some were quite slippy. She had to turn some pieces round or over to get them to stay in her stack.

By sorting the pieces of sea glass by size and by shape (flat and not flat), Nuria found that she could build a tower of ten. Nuria counted back from fifteen and knew she needed five more pieces to make a stack of fifteen. **Why do you think Nuria managed to build a stack with ten layers? Was her stack taller than Amman's? Whose stack had the most layers? How can we make our stone stack taller?**

Amman and Nuria were keen to show the others their towers so they each took a photograph on a tablet. They compared the number of layers in their stacks and shared what they had done. After sharing their ideas, Amman realised that Nuria's idea of sorting the shapes by size before building was a good idea. He was keen to try this out.

Let's reflect!

Provide opportunities to extend the pupils' sorting and classification skills using different objects and resources.

The pupils require experience of building within their setting: ensure that there is a range of stackable materials (not just manufactured 3D objects), both indoors and outdoors, which the pupils might enjoy exploring. Provide sticks, cardboard shoe boxes or commercially produced stacking cups for pupils who are finding stone stacking tricky.

Reflect on whether the children were using / developing their skills:

- Did they share their thinking?
- Were they able to justify their choices?
- Was there evidence of them transferring and applying previous learning?
- Did they use mathematical vocabulary?

Teacher Guide Reference

- Chapter 2 – Number and Number Processes – Number Recognition and Place Value
- Chapter 7 – Measurement

Straws

Experiences and Outcomes

- *MNU 0-11a* – *I have experimented with everyday items as units of measure to investigate and compare sizes and amounts in my environment, sharing my findings with others.*
- **MTH 0-16a** – I enjoy investigating objects and shapes and can sort, describe and be creative with them.

Numeracy and Mathematical Skills

- **Select and communicate processes and solutions** – shares thinking; verbalises or demonstrates thought processes
- **Justify choice of strategy used** – shows and talks through their thinking
- **Uses mathematical vocabulary and notation** – uses developmentally appropriate mathematical vocabulary

Resources

- Straw picture (see Resource sheet – Straws)
- Straws cut into five pieces of varying length – per pair of children (each set may be similar but by being non-identical can result in more solutions and deeper discussion)
- Sticks (as above) as an outdoor alternative

Before they start...

Pupils should be able to recognise and name some common 2D shapes (square, triangle, rectangle)

Let's go!

Display the picture from the IWB and question the pupils about what they see. Ask them **How many straws can you see? Do you think they are all the same length? How many different shapes do you think you could make with these straws? Do you think you will use all the straws to make your shapes? How many sides will your shapes have? The pupils can record their shapes using tablet computers or by making a sketch.** Provide the pupils with their own bundle of five straws. Challenge them to make as many different shapes as they can.

Enabling prompt

Explore and experiment by playing with the straws. **Can you make some shapes?** Encourage the children to look around the classroom or outdoors to see what shapes they can recognise and try to make. Play games such as 'I spy a 2D shape'. Outdoors, provide shape-building experiences by chalking some shapes on the playground for the pupils to place sticks on top of the chalked lines. This can also be done indoors using chalk on black paper or pencil lines on scrap paper.

Extension activity

Can they name any of the shapes? If another straw was added, could they make a six-sided shape?

Encourage them to say what is the same and different to the regular shapes they are familiar with.

Let's check!

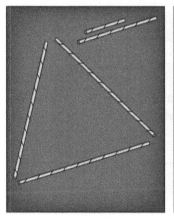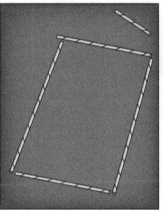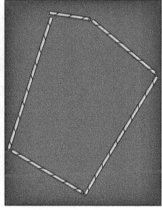

Let's discuss!

The shapes that the pupils make may not be regular-looking shapes and this may challenge some of them. When outdoors, the pupils may choose to find their own sticks – they may compare them for length or just randomly collect them adding some variations to the problem. Isla wanted to make a triangle. She knew it had three sides and three corners; she compared the lengths of the straws and chose three that were almost the same. **Did you make a triangle? Does it look like Isla's one? If not, what does your triangle look like? Remember that not all triangles look the same.**

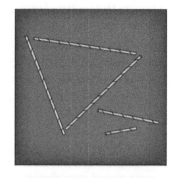

The orientation of Isla's triangle may have challenged some pupils; triangles are frequently represented in one orientation in picture books.

Amman decided he would make a square. He knew that squares have four sides that are the same length. He couldn't find four straws that were exactly the same length.

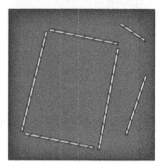

This was the shape he made. He said it looked like a squashed square. **Did anyone make a shape like this? What does it remind you of?**

Finlay took five straws. He didn't measure them. This is the shape he made. He said it looked a bit like a house; he said it didn't look the same as the five-sided shape in the pattern blocks box.

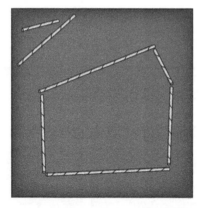

Did you make a five-sided shape? Did it look like Finlay's? Finlay wondered what was different between his shape and the pattern block shape? **Can you think what might be the same and what might be different?**

Let's reflect!

The pupils displayed an understanding of 2D shapes, recognising certain facts related to number of sides and length of sides. It is important to encourage the children to talk about how many sides the shape has and how many corners and this should help build their understanding of the language used in shape. Reflect on whether the pupils recognise and can describe the properties of 2D shapes using mathematical language.

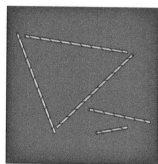

Isla's triangle presents a good opportunity to use transformation to explore the triangle being flipped or rotated. Many young pupils only recognise one orientation and need experience of rotating and flipping shapes.

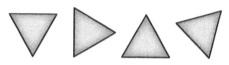

Teacher Guide Reference

- Chapter 7 – Measurement
- Chapter 9 – 2D Shapes and 3D Objects
- Chapter 10 – Directions, Reflections and Symmetry

Using 5

Experiences and Outcomes

- **MTH 0-16a** – I enjoy investigating objects and shapes and can sort, describe and be creative with them.
- **MTH 0-17a** – In movement, games and using technology I can use simple directions and describe positions.
- **MTH 0-19a** – I have had fun creating a range of symmetrical pictures and patterns using a range of media.

Numeracy and Mathematical Skills

- **Interpret questions** – selects the relevant information
- **Select and communicate processes and solutions** – verbalises or demonstrates thought processes
- **Justify choice of strategy used** – shows and talks through their thinking
- **Links mathematical concepts** – transfers learning in one area to another
- **Uses mathematical vocabulary and notation** – uses developmentally appropriate mathematical vocabulary

Resources

- Five wooden craft matchsticks or five large wooden lollipop sticks per child
- Optional – paper and pencil; whiteboard and wipe-clean marker

Before they start...

The children should have experience of:

- exploring shapes using a range of resources, e.g. dough, straws, pipe cleaners, using twigs, drawing with a stick in wet sand
- talking about shapes they have seen (indoors and outdoors) using developmentally appropriate vocabulary
- making pictures (including symmetrical pictures) with materials of their choice (indoors and outdoors)
- rotating objects to explore different perspectives

Let's go!

Show the pupils this image.

Ask: **What do you see? Can you describe what you see? Does it look like anything you know? How many sticks can you see? I wonder what other pictures you could make using five sticks. Can you estimate how many different pictures you might be able to make using five sticks?** This is very much a free exploration task. The children should be able to carry out this task with minimal adult input.

I am going to give you five sticks each and I want you to find as many ways as you can to make different pictures using the five sticks. Do you think you will be able to make some symmetrical pictures?

Enabling prompt

Ask the children how they think they might record the different stick pictures and ensure that they have access to all that they need. For children who are still developing manual dexterity, provide the larger sticks. Encourage them to be creative! Watch and listen to what the children are doing. Are they making links to other mathematical learning? There may be elements of trial and error attached to each picture they make. While the children are exploring, it is worthwhile to ask questions such as, **Can you describe your picture? Can you see any shapes? Can you see shapes within your stick picture?**

Extension activity

Two possible options are included for the Extension activity.

Challenge the children to make symmetrical pictures. **How do you know your picture is symmetrical? Tell me about your picture.**

Provide the option to take one or two more sticks. **I wonder what picture you could make by taking one / two more stick(s). What shapes do you think you could make? Is it easier to make symmetrical pictures using an odd number of sticks or an even number of sticks?**

Let's check!

There are many possible solutions to this task at both levels. Some children will enjoy making pictures whereas others will explore and move one stick at a time to see what happens. It is important to listen for the use of mathematical vocabulary but also links to other areas of learning.

Enabling example

Finlay used his five sticks. He said he had made a zig-zag line. He said it looked like triangles or two mountains and a bit of a mountain. Finlay took a picture using a tablet.

Finlay said his next picture was like the three on a microwave.

He noticed that if he turned his picture round, it looked a bit like the letter m.

Do you agree with Finlay? Where else might we see a number like the one Finlay made? Did you make a number or a shape? Which number did you make with your five sticks? Which shape did you make? Tell me about your shape.

Nuria decided she would make pictures of things from the garden. She made an autumn tree (no leaves) with three branches and a fence. She didn't say anything about shapes or symmetry but she did spot a pattern when making the fence.

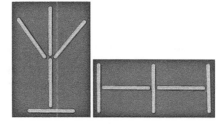

Did you make some pictures like Nuria did? What pictures did you make? Why do you think Nuria said she had made an autumn tree? Was it easy to use all your sticks to make a picture?

Extending examples

Isla made a symmetrical house shape with six sticks. It was made up of a triangle and a square. She also thought it looked a bit like a rocket.

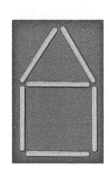

Do you think it looks like a rocket? Can you think of anything else that it might look like? Did you make anything like this? What did you make with your sticks? Did you take six sticks or seven sticks?

Isla wondered if she could make numbers with her sticks. She found she could make digital numbers. Isla enjoyed investigating. She found that by adding one more stick to 5 she could make a 6, and by adding one more stick to a 6 she could make the number 8. Isla thought that this was funny as 5 sticks + 1 stick = 6 sticks, but 6 sticks + 1 stick made the number 8 not 7!

Did anyone explore numbers like Isla did? What numbers did you make? Did you find anything funny like Isla did?

She thought 8 was symmetrical. **Do you agree? How might we check?**

Finlay came to see what Isla was doing. He stood on the opposite side of the table from her. Finlay didn't always see the same numbers as Isla did. **I wonder what he saw instead.** He could see an eight and a five **. . . what do you think he saw instead of a 6?**

Amman chose to use six sticks and made two symmetrical pictures. In his first picture he made two triangles and said it was a superhero mask.

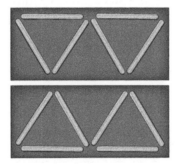

He found when he turned it round it became two mountains.

Is Amman correct when he says it is symmetrical? Do you agree? How might we find out? Did you make a symmetrical picture? What did you make? How many sticks did you use? Amman flipped his shape over to make a different picture. **Did anyone flip their picture over to make something new? What did you make?**

Amman then made a watch like his mum has. He used four sticks to make the square face. He said it was symmetrical.

Do you think Amman's picture is symmetrical? How do you know? Did he use an odd or even number of sticks? Did anyone make a symmetrical picture? How many sticks did you use?

Let's reflect!

Finlay and Nuria were able to describe their pictures and both showed evidence of linking learning. Finlay identified shapes and a number and linked these to his real-life experiences – digital time and mountains. He made a pattern, explored rotating his shape but made no reference to symmetry; however, he did make connections to other areas of learning. Nuria made pictures (a tree and a fence) that were symmetrical but she didn't use any terms to describe symmetry. She said her fence was like a pattern and described it as stick up then stick side, etc. She didn't mention repeating pattern.

Amman and Isla both made symmetrical pictures; they both linked their learning. Isla made and recognised shapes; she moved sticks and explored what happened when she did. She showed deep thinking. Amman was confident in his picture building; he made images that were important to him. He recognised shapes and was excited when he found he could flip his picture over to make another picture. He described the watch as being a square face with the strap coming out two of the sides, providing evidence of mathematical thinking. Amman has transferred his knowledge of the square shape he made to a real-life object, but he is also highlighting his awareness of position by describing the strap as coming out two of the sides.

Some of the ideas that the pupils suggested present ideal teaching opportunities:

- an appropriate time to investigate symmetry when the pupils themselves have made the pictures
- using mirrors to check for symmetry
- sorting some of the pictures into symmetrical and not symmetrical
- Finlay's remark about triangles . . . Is it still a triangle whichever way it is rotated?
- discussing halves in Finlay's remark about 'a bit of a mountain'
- exploring transformation (Isla and Finlay's view of the same object)

Teacher Guide Reference

- Chapter 2 – Numbers and Number Processes – Number Recognition and Place Value
- Chapter 10 – Angles, Symmetry and Transformations

Slide and Roll

Experiences and Outcomes

- **MTH 0-16a** – I enjoy investigating objects and shapes and can sort, describe and be creative with them.

Numeracy and Mathematical Skills

- **Select and communicate processes and solutions** – shares thinking; verbalises or demonstrates thought processes
- **Justify choice of strategy used** – shows and talks through their thinking
- **Use mathematical vocabulary and notation** – uses developmentally appropriate mathematical vocabulary

Resources

A selection of 3D objects: cubes, cuboids, spheres, cylinders, pyramids and cones. Include 'junk' (e.g. empty packaging) and building blocks as well as commercially produced shapes. See Resource sheet – Slide and Roll.

Before they start...

Through play, and life experiences, the pupils will have had opportunities to explore informally the properties of both natural and human-made 3D objects.

The pupils should be able to:

- describe features of 3D objects using appropriate vocabulary, e.g. straight, round, flat, long, pointy
- sort the objects according to their own and others' criteria

Let's go!

I wonder if you can think of ways that we could sort these objects? If no suitable responses are provided, guide the pupils towards thinking about whether the objects will roll, slide or whether they might both roll and slide. Encourage the pupils to think about the properties of an object that enable it to roll or slide but also to think about the surface that they might use to test their predictions. **Why do you think this object will roll? Why do you think this object won't roll? Is it possible for some objects to both slide and roll? Where might be a good place to test your predictions? What makes you think that?**

Provide cards (available in downloads) to help the pupils to organise their sorted shapes.

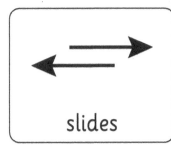

Enabling prompt

Provide opportunities for pupils to play with 3D objects. **What do you notice about this object?** Encourage the pupils to pick the shapes up and feel the curved surfaces and flat faces. Ask **How could we find out if an object rolls or if it slides?**

Play 3D Shape Snap. Shout out: 'curved', 'flat' or 'flat and curved' instead of 'Snap!'.

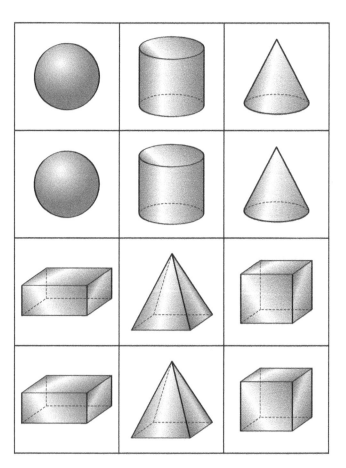

Extension activity

Pupils could show their prediction (P) and findings (F) on the recording sheet available in downloads.

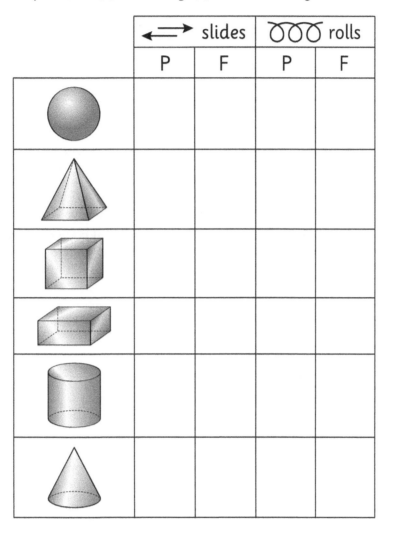

	← → slides		⟅⟅⟅ rolls	
	P	F	P	F
(sphere)				
(pyramid)				
(cube)				
(cuboid)				
(cylinder)				
(cone)				

Let's check!

The pupils should recognise that flat faces and curved surfaces make different motions possible. Shapes that have both curved surfaces and flat faces will be able to make both motions.

Slide: cube, cuboid, pyramid

Roll: cone, sphere, cylinder

Slide and roll: cone, cylinder

Consider, however, that in some cases the material that the object is made from, the size of the object and the type of surface (and elevation) may affect the results, e.g. rough surface may prevent an object from sliding or a tall shape may topple rather than slide.

Let's discuss!

Amman and Isla chose to work on the rug in the book corner; they sorted their objects, tested and organised them into roll, slide and roll and slide. They went to look at what Finlay and Nuria were doing. Finlay and Nuria had chosen to work on the floor in the messy play area. They were surprised to see that Finlay and Isla had sorted their objects differently. **Why do you think the children got different results? Can they all be correct? What could they do to check the results?**

Let's reflect!

Place a selection of 3D objects in a feely bag. Can the pupils describe the shape by touch alone? Ask, **What shape do you think it is? Can you tell me anything that is special about it? Has it got corners? Does it feel flat or curved? Why would this object slide / roll? Where else have you seen shapes that look like these?** This should support visualisation and application of their knowledge of 3D objects.

Teacher Guide Reference

- Chapter 9 – 2D Shapes and 3D Objects

Directions

Experiences and Outcomes

- **MTH 0-17a** – In movement, games and using technology I can use simple directions and describe positions.

Numeracy and Mathematical Skills

- **Interpret questions** – selects relevant information; interprets data
- **Select and communicate processes and solutions** – verbalises or demonstrates thought processes
- **Justify choice of strategy used** – shows and talks through their thinking
- **Use mathematical vocabulary and notation** – uses developmentally appropriate mathematical vocabulary

Resources

- IWB images
- Chalked grid (outdoors) or masking tape grid (indoors)
- Two objects to place in the grid (e.g. stones / toy cars / small soft toy) or pictures
- Hungry Mouse game from the Early Level Assessment Pack
- Blank grids
- Pens / pencils / chalk
- Set of Direction Strips (see downloadable Resource sheet – Directions) for the Extension activity

Before they start...

The pupils should be familiar with the language of position and direction, e.g. forwards / backwards / next to / above / below within play contexts. Some may have an awareness of left and right.

Let's go!

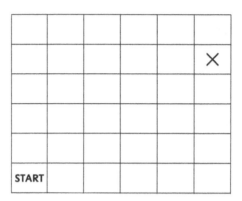

Chalk a grid in the playground, or mark out a masking tape grid in the classroom or hall, as shown here. Say **We are trying to get from *start* to *X*.** (X could be pirate's treasure, the web that the spider wants to get to or any theme that ties in with seasonal or topic-based activities.) **I wonder how we could do this. The rules are that we can only move up, down or sideways.** (The opportunity to discuss 'diagonal' may arise.) **I want you to think about what direction you would move in and how many steps / jumps it would take. How could we record the path?** Give the pupils time to explore the grid and come up with suggestions.

Enabling prompt

Play 'Hungry Mouse' from the Leckie Early Level Assessment Pack or 'location' games (e.g. Hunt the Thimble and Where is Teddy?) in which the pupils have opportunities to use simple directions. They can practice this language as they decide how best to get from one place to another or describe where objects are in relation to each other. Pupils can be supported to develop these skills by placing emphasis on positional vocabulary when tidying up (**Put the box under the table. Please put the book on the shelf next to teddy.**) or when using construction materials, e.g. **Could you please build a tall tower next to the small tower? What colour brick is on top of the green brick?** Playing games such as 'What's the Time Mr Wolf?' can provide practice in counting a given number of steps.

Extension activity

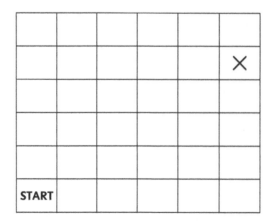

Provide the pupils with a set of Direction Strips.

The feet symbolise the direction and the number of steps or jumps to be taken, e.g. Would four steps / jumps up and four steps / jumps to the right take them to X? **Which strip do you think shows the instructions for getting from Start to X? How will you check if you are right?**

Let's check!

The solutions will depend on how the pupils approach each problem. Some may need to 'walk through the problem' in order to think of solutions. Others may require a smaller grid where fewer steps are required. The pupils can record the direction and number of steps using sticks or twigs when outdoors, and lollipop sticks when indoors. The orientation of the sticks can be changed if necessary. Encourage pupils to keep track of the steps and turns taken at each stage of the journey. Providing a chalked grid or a strip of paper, like Picture 1 will enable pupils to record as they progress and make the data easier to collate at the end.

START								X

Picture 1

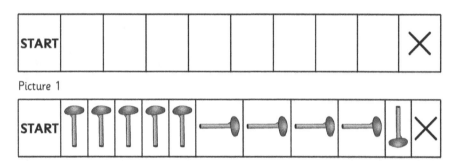

Example of steps taken using pebbles and small sticks.

Let's discuss!

Picture 2

Nuria and Finlay used the chalked grid (above) and some sticks to help them record their solution; they thought that one stick could be used to show one step / jump. **Can you predict how many sticks they would need? Why do you think that? In which direction did they turn? How did you record your solution?**

Once the pupils have discussed the solution, show them Picture 2. **Tell me about your solution. How is this different to Nuria and Finlay's solution? Can you show me your solution? What did you do that was different?**

Amman and Isla chose a Direction Strip.

Finlay said that the strip they had chosen would not finish at X. Amman and Isla disagreed.

They took a blank grid and marked on it to show the steps taken. **Where do you think it will finish? Can you tell or show me why you think that?**

Where have Amman and Isla gone wrong? Do they need more steps or fewer steps to correct their solution? Which way should the extra steps face?

Let's reflect!

Look for pupils who can accurately count the number of steps taken and use one-to-one correspondence to track and record the journey. Check that the pupils are jumping or stepping onto the next square; intervene if the pupils are counting the square they are 'on'. Listen for the use of appropriate positional language to describe the direction taken; some pupils may struggle with positional language but can use the footprints to show their thinking.

Challenge the pupils, in pairs, to find different routes or change the grid layout for variation. Ask them to record their solutions using footprint symbols or their own recording method. **How many different routes can they find? Do all of their solutions involve the same number of steps? Where have you used skills like this before?**

Teacher Guide Reference

- Chapter 10 – Angle, Symmetry and Transformation
- Early Level Maths Assessment Pack. Task 5 Hungry Mouse p. 62

Loose Parts 2 – Making Faces

Experiences and Outcomes

- **MTH 0-19a** – I have had fun creating a range of symmetrical pictures and patterns using a range of media.

Numeracy and Mathematical Skills

- **Select and communicate processes and solutions** – shares thinking
- **Justify choice of strategy used** – shows and talks through their thinking
- **Use mathematical vocabulary and notation** – uses developmentally appropriate mathematical vocabulary

Resources

Set up an outdoor shop or have tubs / boxes, etc. containing **a selection of loose parts**, e.g. pine cones, leaves, selection of sticks and twigs, chestnuts, moss, bottle tops, variety of pebbles, stones.

Before they start...

The pupils should be able to:

- recognise that objects or pictures that are symmetrical have a 'sameness' on two sides
- have experience of creating symmetrical pictures, e.g., by painting on one half of a piece of paper and folding to create a mirror image

Let's go!

Choose some items from the Outdoor shop that you think you could use to make a symmetrical face picture. You may return to the shop if you require more items to finish your picture. Once you have finished your picture, take a photograph of it using a tablet.

Show the pupils these pictures as a stimulus. Ask, **Tell me about this picture / these pictures. What loose parts have been used to make the faces? What do you think the pine cones were used for? What would a face picture look like if it was symmetrical? Do you think either / both of these pictures are symmetrical? Can you tell me why you think that? What do you think you might use to make your face?**

Enabling prompt

Check that the pupils understand the challenge and know what 'symmetrical' means. Chalk or draw a line of symmetry to assist any pupils who are uncertain about making their picture symmetrical. Show them how to place a few items such as bottle tops on either side of the line of symmetry. Start by placing an item on one side of the line and ask the pupil to place theirs on the other side of the line, reflecting yours.

Support them as they select their items: **What are you going to choose? What might you use for eyes? Do you think we need to find two the same? Why? How many of * do you think you will need? What do you think you will need to do next to make the face symmetrical?**

Extension activity

Have a pile of items ready. Challenge the pupils to work in pairs to create a symmetrical face. Each child takes turns to pick an object and put it on their side of the face. Can they work together to make a symmetrical face?

Let's check!

This is a versatile open-ended task that has many possible solutions. You should check to see if the pupil's picture is symmetrical by shape and then by colour.

Let's discuss!

Amman remembered making symmetrical pictures with shapes. He took two bottle tops for eyes; he chose two orange ones. He chose a small stick for the mouth and a leaf for the nose. He used four items. He checked to see if his picture was symmetrical by placing a straw down the middle of his face picture; he used this to represent the fold line that he might have used if he was using paper. He had an eye on each side, opposite each other, the leaf was under the straw and the stick mouth seemed to cross the line and have the same amount of stick on each side of the straw. **Can you tell me what you did to check if your picture was symmetrical?**

Nuria went straight to the shop and grabbed some items. She wanted hair and ears on her face as well as eyes, a nose and a mouth. She sorted through the mixture of items she had taken – she had too many things! Out of her collection, she chose six leaves for the hair and two pine cones for the ears, two bottle tops for eyes, six pebbles for a mouth and a stick for a nose. Nuria counted up the items she had used; it was 17 items. Nuria said her picture was symmetrical; she had used an odd number of items. **Is this possible? What did you find out when you made your picture?**

Let's reflect!

The children demonstrated an understanding of how to make a symmetrical picture and used appropriate mathematical language – bottle tops were **opposite** each other, the stick was in the **middle**. Amman was able to think about his previous learning about symmetry and transfer it to his picture. Nuria may want to explore making a symmetrical face using an even number of items. Ask the pupils to look at the pictures they made – **Can you describe what you did to make your picture symmetrical? Did you use a mirror to check? Can you tell me why you picked the items that you used?**

Teacher Guide Reference

- Chapter 10 – Angle, Symmetry and Transformation

Garden Flowers

Experiences and Outcomes

- *MNU 0-20b* – *I can match objects, and sort, using my own and other's criteria, sharing my ideas with others.*

Numeracy and Mathematical Skills

- **Interpret questions** – interprets data
- **Select and communicate processes and solutions** – explains choice of process; shares thinking; verbalises thought processes
- **Justify choice of strategy used** – shows and talks through their thinking
- **Use mathematical vocabulary and notation** – uses developmentally appropriate mathematical vocabulary

Resources

Photocopiable picture cards (see Resource sheet - Garden Flowers)

Flowers 1

Flowers 2

One picture card between two pupils or display on IWB.

Before they start...

Pupils should be able to:

- identify colours
- use vocabulary related to shape and size

Let's go!

Flowers 1

Flowers 2

Display the picture card on an IWB or hand out one picture card per pair of pupils. Use Flowers 1 for the main activity and Flowers 2 for the Extension activity. Provide sufficient time for the pupils to discuss the flower pictures in pairs and decide which one doesn't belong. Challenge each pair to justify their answer.

Enabling prompt

For pupils who are struggling to get started, guide them to look closely at the pictures. **What can you see? Are the pictures all the same? Can you see things that are the same? Can you see something that is different?** Initially, the pupils may only see one solution.

Extension activity

The Extension activity provides more challenge due to the flowers all being the same colour. **What can you see? Are the pictures all the same? Look carefully at the four pictures – what can you see that is the same or different?**

Let's check!

There are lots of different solutions; much depends on what the children see, their background knowledge of the content and their understanding of what the picture portrays. The pupils may focus on the colour of the petals, the size of the flower, what they can see in the background and the number of flowers. Pupils enjoy searching for small details that are hidden in a 'sea of noise'. Flowers 1 has more obvious differences whereas Flowers 2 is slightly more challenging as all the flowers are pink.

Let's discuss!

These solutions were found by Amman, Isla, Finlay and Nuria. I wonder if you thought of something similar.

Enabling prompt

Amman and his partner said ○ (yellow) was different. They said that the picture was of one flower. They said the rest of the pictures had lots of flowers.

Isla and her partner said ● (blue) was different. They said that the flowers were pink and the rest were purple.

Finlay and his partner said ● (green) was different. They said that there were lots of flowers in this picture but the others didn't have as much.

Nuria and her partner said ● (green) was different. They said that that it was a long flower shape and the rest were roundish.

Did anyone make the same choice as Amman / Isla / Finlay / Nuria? Do you think their solutions were correct? Is this the same as your solution? What did you see? Can you tell me something that you thought was different?

Extension activity

Amman and his partner said **c** was different. They said that the small petals made up a big flower that was like a big circle and the others were not that shape.

Isla and her partner said **d** was different. They said that the middle of the flower was yellow and the others weren't.

Finlay and his partner said **a** was different. They said this flower had striped petals and the others didn't.

Nuria and her partner said **b** was different. They said that that the long flower was made up of lots of tiny petals.

Did anyone think of the same solutions as Amman / Isla / Finlay / Nuria? Which picture did you choose? Can you tell me why you thought that one was different? Did anyone think of a different solution?

Let's reflect!

The children noticed a lot of information about the pictures and were able to interpret the pictures with minimal practitioner input. They were mostly able to justify their thinking. They picked up differences relating to:

- the colours of the petals / flowers
- the difference in petal size
- the shape of the flowers
- the number of flowers (more and less)

The differences that pupils notice could be expanded upon in a plenary. **I wonder if there are any other differences that you can spot?** It is always worthwhile asking this once everyone has presented their solutions, as pupils may think of other solutions when listening to what their peers have to say.

A 'which one doesn't belong' task:

- is open ended
- provides a forum for pupils to justify their ideas
- provides an opportunity for pupils to use their mathematical language in context

Teacher Guide Reference

- Chapter 9 – 2D Shapes and 3D Objects
- Chapter 11 – Data and Analysis

Sort Me Out!

Experiences and Outcomes

- *MNU 0-20b* – *I can match objects, and sort, using my own and other's criteria, sharing my ideas with others.*

Numeracy and Mathematical Skills

- **Select and communicate processes and solutions** – shares thinking
- **Justify choice of strategy used** – shows and talks through their thinking
- **Uses mathematical vocabulary and notation** – uses developmentally appropriate mathematical vocabulary

Resources

- copy of the image shown per pair of children
- selection of wooden, card or plastic letters and numbers as shown

t   r   q   s
        g
c   A   a   4   6

10   d   q   8   V

3   5   K   7   B

l   2   m   0

Before they start...

The pupils should be able to:

- distinguish between letters and numbers
- use and understand developmentally appropriate mathematical vocabulary, e.g. curved, straight, line
- sort objects using their own criteria

Let's go!

Display the image of the various numbers and letters. Explain to the pupils that after tidy-up time, these objects were found under the tables. Mrs Jones, the class teacher, needs help to decide what should be done so that they might be put away in the correct places. **What you can see? What is the same? What is different? How could you sort them?** Note: the letters and numbers selected may be changed to suit the group.

Enabling prompt

Provide the pupils with cut-out, tactile letters and numbers. Restrict the quantity of items provided if required. **Tell me what you see. What is the same? What is different? Which objects have straight / curved lines? How can this help us to sort them?**

Extension activity

Increase the number of symbols to be sorted. Can the pupils use two criteria to sort the letters and numbers? For example, letters that are symmetrical.

Let's check!

The objects can be sorted in a number of ways:

- letters and numbers
- capital letters, lower-case letters and numbers
- curved lines, straight lines or curved and straight lines (leading to display in a Venn diagram)
- the way that a letter or number is formed when written, e.g. with one continuous 'action' (r) or more than one 'action', i.e. when the pencil or pen is lifted from the page between actions (B)
- symmetrical / not symmetrical

Let's discuss!

Isla decided to sort the objects into two bundles – letters and numbers – because she knew that Mrs Jones had a tray for letters and a tray for numbers. **How did you sort the objects?** Isla finished quickly so she took out her whiteboard and marker and thought she would practise writing the letters and numbers. She found that as she wrote some letters and numbers, she didn't need to lift her pen but for others she had to lift her pen two or three times. **I wonder which letters Isla needed to lift her pen for? Do you think she would be able to write m or 8 without lifting her pen?** If appropriate, challenge pupils to sort the numbers and letters using these criteria.

Amman looked carefully at the picture. He saw capital letters and lower-case letters – 'A' for the start of his name and also 'a' within his name. He also saw numbers. He decided that he would sort the objects into three bundles – capital letters, lower-case letters and numbers. **Who else sorted the objects into capital letters, lower-case letters and numbers? Did everyone make the same three sets? Can you all be correct?**

Nuria looked at the shapes of the letters and numbers. She said that some had only curved lines (0, 8, 3, 6) some had only straight lines (V, 7, A, 1, K) but others had both curved and straight lines (t, r, g, d, 5, 10, 9, s, c, B, a, q, 2, m). She said she would sort them into three bundles: straight lines only, straight and curved lines and curved lines only. She said that the straight *and* curved lines bundle should be set out in the middle of the other two bundles. **Do you think Nuria is correct? Did you sort the objects in this way? How did you sort them?**

Finlay noticed some letters and numbers were symmetrical. He thought that they should be sorted into symmetrical (A, V, 8, K, B, I, 0, c, 10, 3) and not symmetrical (t, r, g, 9, s, a, 4, 6, d, q, 5, 7, 2, m). Finlay looked at the picture again and realised that it was possible to sort the objects another way – this time into 4 groups – capital letters, lower-case letters, single-digit numbers and two-digit numbers. **Do you agree with Finlay? Can you think of any other numbers that could be in the two-digit group?**

Let's reflect!

The pupils are being asked to sort by looking at the complete set. Some might find this difficult and it may be necessary to offer individual tactile letters and numbers.

As pupils sort, they should be encouraged to look closely at the features of letters and numbers and to apply other numerical and mathematical knowledge that they may have to the problem. Can they describe the numbers and letters using the terms curved and straight? Many young children have difficulty visualising and forming letters and numbers and this type of activity spotlights similarities and differences. Ask the pupils to tell you what is the same and what is different. Making the letters and numbers by rolling out dough can help consolidate understanding. There should be evidence of creativity and justification.

All of the pupils recognised that there were both letters and numbers displayed on the image and sorted the symbols accordingly. Nuria's solution would naturally lead to exploring the representation of data in a Venn diagram. Finlay's response provides a stimulus for exploring symmetry and for discussing single-digit and two-digit numbers.

Teacher Guide Reference

- Chapter 11 – Data and Analysis